THE FUTURE OF WORK

THE FUTURE OF WORK

ROBOTS, AI, AND AUTOMATION

Darrell M. West

Brookings Institution Press
Washington, D.C.

THE BROOKINGS INSTITUTION
1775 Massachusetts Avenue, N.W., Washington, D.C. 20036
www.brookings.edu

Library of Congress Cataloging-in-Publication Data
Names: West, Darrell M., 1954– author.
Title: The future of work : robots, AI, and automation / Darrell M. West.
Description: Washington, D.C. : Brookings Institution Press, [2018] | Includes bibliographical references and index.
Identifiers: LCCN 2018012410 (print) | LCCN 2018021828 (ebook) | ISBN 9780815732945 (ebook) | ISBN 9780815732938 (hardcover : alk. paper)
Subjects: LCSH: Work. | Robots. | Automation. | Artificial intelligence. | Social contract.
Classification: LCC HD4855 (ebook) | LCC HD4855 .W47 2018 (print) | DDC 306.3/60112—dc23
LC record available at https://lccn.loc.gov/2018012410

First printing in paperback, 2019 (ISBN 9780815737865)

9 8 7 6 5 4 3 2 1

Typeset in ITC Legacy Serif

Composition by Elliott Beard

To Jenny Lu Mallamo,
Hillary Schaub, and
Liz Valentini

The best assistants an author can ever have

CONTENTS

PREFACE

I REALIZED SOMETHING noteworthy was happening when my assistant Hillary came to me with an unusual experience. I had asked her to reschedule an appointment and she had emailed Amy, the personal assistant of the individual with whom I was to meet. Amy was amazingly prompt in her follow-up, and when she did not get a response from Hillary over the weekend, she emailed my assistant multiple times trying to find a date that would work.

It was only at this point Hillary noticed Amy was a "virtual assistant." Working for an artificial intelligence (AI) firm that schedules meetings, Amy performed the tasks of a human assistant who read emails, discerned intent, and came up with a relevant response. Other than the AI title buried in her signature line, there was nothing in the exchange that would lead anyone to conclude the correspondent was virtual other than her incredible persistence over a weekend.

In reflecting on this experience, I realized that a digital assistant trained in intelligent response is not a futuristic vision. Rather, it is a current reality that performs quite well. This and other automated tools no longer are alone at the cutting edge of technology. Rather, robots, AI, virtual reality, autonomous vehicles, facial recognition algorithms, drones, and mobile sensors are altering numerous sectors and leading us to an automated society.

In this book, I explore the impact of these emerging technologies on work, education, politics, and public policy. If companies need fewer workers as a result of automation and robotics, but most societal benefits are delivered through full-time jobs, how are people outside the workforce for a lengthy period of time going to get income, health care, and retirement pensions? In this situation, it is important to rethink work and move toward lifetime learning so that people are trained for a world of dislocation. There are reforms in the social contract that would ease the transition difficulties, but it is not clear the U.S. political system is up to the task of adopting relevant policies. If leaders don't make the right choices, developed nations could end up facing serious economic and political disruptions.

The plan of this book is as follows. Chapter 1 looks at the growing use of robots. These devices are increasing in sophistication and dropping in price. In the process, they are transforming commerce and ushering in new business models. The reality of a large workforce with full-time jobs and benefits is giving way to an economy based on temporary employees, partial or no benefits, and widespread automation.

Chapter 2 reviews advances in AI, machine learning, facial recognition, driverless cars, drones, and virtual reality. Rather than requiring human intervention, improvements in software design make it possible to perform complex tasks using sophisticated algorithms. The result is an increase in economic activity but limited full-time employment opportunities other than for workers such as coders, computer experts, designers, and data scientists. These innovations are changing the way companies operate and altering the relationship between managers and employees.

Chapter 3 explores the growing reliance on sensors and the emerging network known as the Internet of Things. Digital devices are spreading in number and enabling important advances in finance, health care, transportation, public safety, and resource management. With the coming 5G network, homes and businesses will be connected through high-speed broadband, and that will make possible a dramatic expansion of digital services that will transform commerce and communications.

Chapter 4 argues that at a time of accelerating technology, we need to consider the ramifications for the labor force and rethink the concept of work itself. In thinking about the future, we must broaden the notion of employment to include volunteering, parenting, and mentoring, and also pay greater attention to leisure-time activities. New forms of identity will be possible when the "job" no longer defines people's personal meaning and those in the workforce have time to engage in hobbies, personal interests, and community projects.

Chapter 5 examines the need for a new social contract and the implications of changing employment for income provision, health care benefits, and retirement support. Right now, many social benefits are tied to jobs, which limits the benefits to those who are fully employed. However, as the business model changes, more people will find themselves underemployed or in positions that don't provide benefits. In this situation, social benefits will need to become portable and flexible as workers move in and out of jobs. Unless there are innovative service delivery models, there may arise a large and permanent underclass that does not receive job benefits and is trapped in poverty.

Chapter 6 calls for lifetime education to help workers and employers deal better with digital disruptions. The world is going through an extraordinary period of large-scale change driven by technology innovation and changing business models. Outsourcing has become prevalent, and the emerging economy necessitates education and training programs throughout adulthood. People will need to acquire additional skills in order to remain competitive in the twenty-first-century economy.

Chapter 7 asks whether American politics is up to the challenge of a transition to a digital economy. It is difficult for business and government to redefine work, develop a new social contract, and help people gain the skills they will need. Society is fragmented, governance systems are polarized, news coverage is not very substantive, and it is hard for people to have meaningful conversations about how to reimagine the social contract. Figuring out ways to facilitate productive discussions and address the resulting political tensions will be a major challenge in coming decades.

Chapter 8 summarizes the major recommendations of the book. I argue that to cope with automation, we need to undertake a number of economic and political reforms. These include enacting paid family and medical leave, expanding the earned income tax credit, building a Republic 2.0 with political institutions capable of dealing with economic dislocations, passing universal voting to reduce political polarization, abolishing the Electoral College, reforming campaign finance, and adopting a solidarity tax to fund needed social programs.

An early version of this project was presented in 2015 through my Brookings Institution paper titled "What Happens If Robots Take the Jobs? The Impact of Emerging Technologies on Employment and Public Policy." In it, I looked at the accelerating nature of technology innovation and the ramifications for employment, workforce development, and public policy. I appreciate the help of Gisele Huff, Gerald Huff, and Jerry Hume in supporting that project. This book also draws on publications I wrote exploring AI, the Internet of Things, driverless cars, digital education, mobile technology, smart transportation, megachange, news media, and inequality.

I wish to thank several people for their help with this book. I am indebted to Jack Karsten, Karin Rosnizeck, Jake Schneider, Nicole Turner-Lee, and Tom Wheeler for conversations about the book. Grace Gilberg, Jack Karsten, Hillary Schaub, and Kristjan Tomasson provided valuable research assistance on this project. External reviewers Kevin Desouza of Arizona State University and Dipayan Ghosh of Harvard University and the New America Foundation provided helpful suggestions on the manuscript. A number of individuals at the Brookings Institution Press deserve a special thank you. Press director William Finan and assistant director and sales manager Yelba Quinn provided invaluable counsel on the title and marketing for the book. Janet Walker and Elliott Beard deserve a big thank you for overseeing production, and Marjorie Pannell did an excellent job editing the volume. None of these individuals is responsible for the interpretations, which are mine alone.

PART I
ACCELERATING INNOVATION

ONE
ROBOTS

RESTAURANT EXECUTIVES ACROSS the United States are reacting to tight labor markets by introducing automated tablets that transmit food orders. Rather than use the services of wait staff, customers place orders through mobile screens. Andrew Puzder, former CEO of CKE Restaurants, the parent company of Hardee's, praised digital devices over human workers. Referring to the former, he said, "They're always polite, they always upsell, they never take a vacation, they never show up late, there's never a slip-and-fall, or an age, sex or race discrimination case."[1] Noting labor requests for a higher minimum wage, writer Eric Boehm of Watchdog.org opined that "a computer kiosk doesn't need to be paid $15 an hour to take orders."[2]

McDonald's, meanwhile, has announced plans to install "digital ordering kiosks" in place of cashiers at 2,500 of its American restaurants and mobile ordering at 14,000 of its stores. Based on these technologies, market analysts in 2017 raised their 2018 growth projections for the firm from 2 percent to 3 percent. McDonald's believes that digital tools cut costs, improve productivity, and reduce the chain's reliance on human employees. The corporation's officers predicted that the new technologies would lift the company's stock price by 17.5 percent in 2018.[3]

These restaurant firms are not alone in embracing digital automation. Amazon is replacing cashiers in its new storefront locations. Rather than employ humans to scan purchases and generate a bill, Amazon Go "allows customers [to] check in to the store using a smartphone app and walk out with what they need." Sensors track items that people want to buy and charge their accounts.[4] This innovation is significant for overall employment because retail clerks and cashiers constitute 6 percent of the U.S. workforce, or about 8 million workers in all.[5]

In addition, Amazon has expanded rapidly into robots in its distribution warehouses. It has deployed around 55,000 Kiva robots, up from 30,000 in 2016, with many more expected in the future.[6] According to Marc Wulfraat of the consulting company MWPVL International, "Picking is the biggest labor cost in most e-commerce distribution centers, and among the least automated. Swapping in robots could cut the labor cost of fulfilling online orders by a fifth."[7] The virtue of robots is that they can move heavy racks, locate products for shipping, and place the relevant items in a box, all without human intervention. As robots learn how to handle new objects in the warehouse, each "shares what it learns with a hive mind in the cloud" and helps other automated machines locate items.[8]

Truck driving long has been a well-paying job for high school graduates. This occupation does not require a college degree and is an attractive entry-level position for those not seeking higher education. According to Brookings economist Alice Rivlin, in 2016, "There were 1.7 million heavy and tractor-trailer truck drivers, with a median annual wage of $43,590; 859,000 light-truck and delivery workers, who earned $34,700; and 426,000 driver/sales workers, who earned $28,449. So the rough estimate would be that driverless deliveries would put at least 2.5 million drivers out of work."[9]

As illustrated by these examples, the list of emerging technologies grows every day. Robots, autonomous vehicles, virtual reality, artificial intelligence (AI), machine learning, drones, and the Internet of Things are moving ahead rapidly and transforming the way businesses operate and how people earn their livelihoods. For millions who work in occupations such as food service, retail sales, and truck driving, machines are replacing their jobs. There already is evidence of this happening with blue-collar jobs, but the impact is starting to be felt by the white-collar workforce as well.

In this book, I analyze several aspects of the technology revolution. First I review developments in robotics, AI, and sensors associated with the Internet of Things, and show how they are transforming business. I then look at how these digital technologies are redefining jobs and altering financial models. After that, I examine how the social contract should be reconfigured to cope with these transformations and the manner in which health care, income, and retirement benefits are provided. Finally, I discuss whether our political processes in a polarized society are up to the task of handling the transition to a digital economy and how we can cope with an automated society.

This is not the first time people have encountered megachange, whether of a social, economic, political, or technological variety.[10] One hundred years ago the United States (and other countries) made the transition from an agrarian to an industrial economy. It took several decades to work through the resulting transformations in business models, employment, and social policy, but leaders rose to the challenge of dealing with those disruptions.

Today, as the United States moves from an industrial to a digital economy, poor governance poses a serious barrier to expanding the definition of jobs, revising the social contract, and extending models of lifetime learning. With the current political dysfunction in the United States, the high levels of economic inequality, polarized media coverage, and societal divisions, it is not clear that economic and political leaders can resolve the anxieties and dislocations associated with technology-induced disruption. Unless there is more effective governance, the process of conflict resolution will prove quite contentious over the next few decades and could undermine democratic systems of government. As I note in the concluding chapter, we need fundamental economic and political reforms to deal with these challenges and make sure we have a smooth adjustment to the emerging economy.

THE GROWING USE OF ROBOTS

The use of robots is expanding around the world. About 5.4 million were sold in 2015, and that number doubled in 2016 to more than 10 million units.[11] The top applications were in manufacturing, construction, rescue operations, and personal security.

The use of industrial robots deployed in factories has also expanded. Figure 1-1 shows the number of these devices in operation globally; as is evident from the figure there has been a substantial increase in the past few years. In 2013, for example, an estimated 1.2 million industrial robots were in use. This figure rose to around 1.5 million in 2014 and increased to 1.9 million in 2017.[12] Japan has the most, at 306,700, followed by North America (237,400), China (182,300), South Korea (175,600), and Germany (175,200). Overall, robotics is expected to grow from a $15 billion to a $67 billion sector by 2025.[13]

According to an RBC Global Asset Management study, the reason for this expanded usage is that the costs of robots have fallen substantially. It used to be that the "high costs of industrial robots restricted their use to few high-wage industries like the auto industry. However, in recent years, the average costs of robots have fallen, and in a number of key industries in Asia, the cost of robots and the unit costs of low-wage labor are converging. . . . Robots now represent a viable alternative to labor."[14] To illustrate this point, a warehouse in California that introduced robots at a cost of $30,000 to $40,000 per unit found that robots could "handle 30% to 50% of the items the facility ships each day, in about half the time it takes a human worker."[15]

A CEO of a top technology firm explained the new financial model facilitating robotics and its effects on the employment prospects of lower-skilled workers: "We will soon launch a robot that can perform tasks currently done by people with a high school education or less. The robot will only cost $20,000. We're not the only ones; our competitors across the world are working on similar projects. When these cheap, efficient and reliable robots become commonplace, I have no

Figure 1-1 Number of Industrial Robots around the World

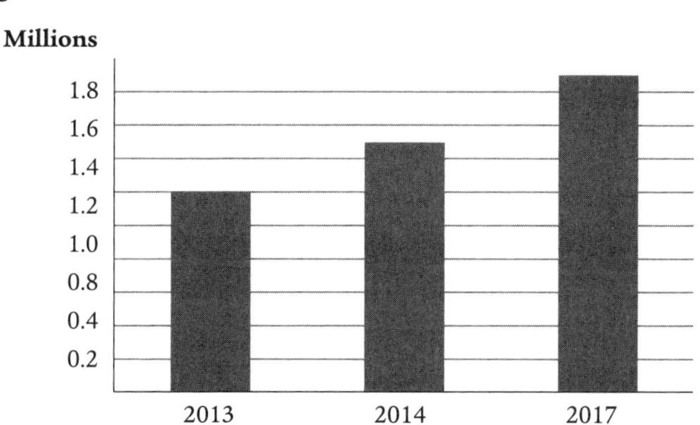

Source: Alison Sander and Meldon Wolfgang, "The Rise of Robotics," Boston Consulting Group, August 27, 2014. The 2017 numbers are projected figures.

idea what jobs will be given to people who don't have skills above a high school level."[16]

Other executives also emphasize the declining cost of robots as a key feature in their adoption decisions. Factory owner Joe McGillivray runs a company called Dynamic that manufactures plastic molds and metal parts. In his factory, where it once took four people to operate a press making the molds, he purchased a robot for $35,000 that was effective at doing their jobs. It worked well and was easy to reprogram for work tasks.[17]

The Hudson's Bay Company, meanwhile, has deployed robots in its distribution center and found positive results. According to Erik Caldwell, senior vice president of supply chain and digital operations, "This thing could run 24 hours a day. They don't get sick; they don't smoke."[18] Combined with low cost, those qualities give robots important advantages over human workers.

With recent efforts in the United States and elsewhere to increase the minimum wage and provide benefits for human workers, the compensation differential between a robot and a human has dropped even further. A paper by the economists Grace Lordan and David Neumark, for example, found that "increasing the minimum wage decreases significantly the share of automatable employment held by low-skilled workers, and increases the likelihood that low-skilled workers in automatable jobs become unemployed."[19] This view was echoed by Wendy's COO Bob Wright, who noted, "We've hit the point where labor-wage rates are now making automation of those tasks make a lot more sense."[20]

These are just a few of the ways in which robotic devices are altering businesses. As a sign of their growing sophistication, the Defense Advanced Research Projects Agency held a competition for robots that could perform effectively in hazardous environments. Robots were given eight tasks, including "driving a vehicle, opening a door, operating a portable drill, turning a valve and climbing stairs."[21] The goal was to have equipment that could operate in damaged nuclear reactors or at disaster scenes considered too dangerous for humans.

In this competition, a team from the Korea Advanced Institute of Science and Technology won the $2 million first-place prize by building a robot called Hubo that completed each of these tasks without falling down. The device was five feet, seven inches tall and weighed 200 pounds. With two arms, two legs, and a head featuring a LiDAR camera, it could scan its surroundings as it maneuvered around obstacles in a search-and-rescue mission.[22]

Robotization is very popular in China. Farmers there are deploying "nanny robots" to monitor the health of their chickens. Using mechanized machines equipped with the latest sensors, these devices identify and isolate "feverish or immobile birds from their cages to protect the rest of the brood and keep sick birds and their eggs from reaching kitchen tables." Firms such as the Charoen Pokphand Group use eighteen "automatons" to make sure that bird flu does not break out. With the poultry sector generating $100.7 billion in revenue, companies see technology as a way to promote food safety while also improving business efficiency.[23]

Some Chinese factories are operated largely by robots.[24] In Hangzhou, for example, a Ford assembly plant utilizes 2,800 workers and 650 industrial robots that automate car production.[25] Tasks such as welding and painting have been automated, and applying protective sealants is expected to be undertaken by robots in 2018. This is part of a massive expansion in industrial robotics in China. New plants have opened in Shanghai, Wuhan, and elsewhere around the country.

A factory in Dongguan City, China, is operating almost exclusively with robots. The facility, run by the Changying Precision Technology Company, "has automated production lines that use robotic arms to produce parts for cell phones. The factory also has automated machining equipment, autonomous transport trucks, and other automated equipment in the warehouse."[26] A handful of human workers oversees the production line, while sixty robots perform the tasks that used to require 650 employees. The robots have increased the annual production from 8,000 to 21,000 phones and reduced the defect rate from 25 percent to 5 percent.

Not to be outdone, Foxconn, the Chinese company that makes Apple iPhones, has established a goal of "30 percent automation at its factories by 2020." Using 10,000 "Foxbots," the firm already has eliminated 60,000 human jobs through robots and automated operations.[27] It is altering the workplace by deploying these machines and engineering new efficiencies in the manufacturing process.

In Japan, Henn-na Hotel in Nagasaki Prefecture uses robots to check in guests and escort them to their rooms. The robotic receptionist speaks Japanese or English, depending on the preference of the guest. It can set up reservations for guests, take them to their rooms, and adjust the accommodation's temperature. Within the room, guests can use voice commands to alter the lighting and ask questions regarding the time or weather.[28]

Finally, automated devices are improving people's educational experiences. A ten-year-old American schoolgirl named Peyton Walton uses a "virtual self" robot to attend classes while she is receiving cancer therapy 250 miles away from her school. The robot has an iPad screen in the classroom that allows Peyton to "join in the day's activities, talk to teachers and navigate her classroom, [with] her face showing in real time" on the computer screen. The two-way communication interface enables the young girl to continue her education while undergoing a course of radiation therapy and helps her maintain some normalcy and classroom connection while receiving medical treatment.[29] These are just a few of the ways that automated processes are transforming a variety of sectors.

ROBOTS THAT LEARN AND ADAPT

Robots used to be limited to executing mechanical, repetitive activities. Factory tools that performed one task really well were commonplace, and they were very effective at relieving humans of day-to-day drudgery. There was no need to waste human time on activities that robots could perform efficiently and effectively.

Today's robots and automated machines, however, go far beyond repetitive tasks. They take on sophisticated work and adjust their decisions as they perform various activities. For example, current machines learn from the experiences of other devices. Autonomous vehicles can compile information on the roadway and pinpoint with great precision where potholes or traffic detours exist. Once they have that information, they send it in real time to other vehicles that are on the road and inform them of upcoming obstacles. Those cars then adjust and make use of the new data.

Machines that sense and learn are capable of much greater sophistication than those that cannot adjust as they perform a fixed set of tasks. Indeed, it is the capacity for self-learning that distinguishes today's robots from those of previous generations. They can undertake specific tasks and adjust their performance as they gain knowledge in the process.

Some automated machines even are capable of creative activities. The anthropologist Eitan Wilf of Hebrew University, Jerusalem, who studies improvisational music robots, has seen a "jazz-improvising humanoid robot marimba player" that can interpret the musical context and respond creatively to improvisations on the part of other performers.[30] Designers have introduced it into a jazz band and the robot ad libs seamlessly with the rest of the group. If someone were listening just to the sound, that person could not discern the robot from the human performer.

Seeking to improve its warehouse operations, Amazon has expanded the use of robots that can "autonomously grab items from a shelf and place them in a tub." It has organized competitions in which robots developed by outside firms demonstrate various kinds of competencies. During one recent event, a Berlin robot successfully completed ten of the twelve assigned tasks.[31] Its automated devices matched orders with products and took them to a mailing unit for customer transmission. That capacity eliminates the need for humans to walk around warehouses to fill orders.

Ahti Heinla has built a twenty-five-pound delivery robot called Starship that combines mobility, wireless technology, and GPS mapping software to deliver goods autonomously to customers. The company has targeted "bakeries, groceries, couriers, and other businesses looking to deliver within a 10-square-mile area."[32] The robot is being tested in twenty-two European and American cities and so far has performed well in replacing human delivery.

Robots have also moved into the security guard business. The mixed use development Washington Harbour in the District of Columbia has a robot named "Steve" that is five feet tall and features video cameras with a 360-degree view. It navigates the mall area around the harbor and collects a variety of information. According to its owners, "We use thermal imaging to detect potential fires. We note license plates to identify suspicious cars that linger for suspicious lengths of time. We take photos and video that our human clients can use to assess suspicious activities." Its developers claim the robotic security guards have many advantages: "We do not tire. We do not take sick days. We do not unionize. We cost $7 an hour."[33]

SOCIAL ROBOTS

There furthermore has been a rise in the sales of "social robots," which provide companionship. According to designers, "A key factor in a robot's ability to be social is [its] ability to correctly understand and respond to people's speech and the underlying context or emotion."[34] Early versions of companionable social robots, in the form of small robotic pets, appeared at the turn of the twenty-first century. Gradually they have become more sophisticated and humanoid, and have even been tested for interactive ability to improve their users' emotional health. In senior citizens centers in Italy and Sweden, pilot projects have tested robots in the care of 160 seniors and found they were able to "assuage loneliness and isolation and reduce health-care staff."[35] Mechanical aides also "pick up groceries and take out the trash" for older people. These devices ease the concerns of family members and allow elderly patients to receive high-quality care.

Other firms are experimenting with indoor drones that help the elderly or disabled retrieve distant objects without moving. For example, if patients need medication stored in the bathroom, experimental drones will find and deliver the medication. There also are new developments in "intelligent walkers, smart pendants that track falls and 'wandering' room and home sensors that monitor health status, balancing aids, virtual and robotic electronic companions, and even drones."[36]

Smart baby monitors are assisting with child care. Mattel's programmable device known as Aristotle, a sort of "Echo for Junior," can, according to its developers, "help purchase diapers, read bedtime stories, soothe infants back to sleep, and teach toddlers foreign words."[37] Such virtual assistants combine high-definition cameras with voice-activated features to perform key tasks and interact with the young child. Aristotle, voiced by a woman, can play games or answer questions through an interactive interface.

Some families are using Amazon's Echo device known as Alexa to "coparent" children. The writer Rachel Botsman allows her three-year-old daughter, Grace, to play with Alexa and pose a variety of questions about the weather, music, and math. After some time spent getting familiar with the device, she found her daughter treating Alexa as a daily companion that could be trusted as an information source. As an illustration, Alexa helps Grace decide what to wear each day.[38]

Other people are using a robot known as Nao to deal with stress. In a project called "Stress Game," researchers Thi-Hai-Ha Dang and Adriana Tapus subject participants to a board game in which they have to collect as many hand objects as they can. During the test, stress is altered through game difficulty and audible signals when errors are made. The individuals are wired to a heart monitor so that Nao can understand their stress levels. When the robot feels human stress increasing, it provides coaching designed to defuse the tension. In this way, the "robot with personality" is able to provide dynamic feedback and help people deal with difficult encounters.[39]

SEX ROBOTS

In the early days of cable television and the internet, the most profitable sector was pornography. Customers were willing to pay substantial money for access to X-rated videos and websites with interactive chatrooms. Without leaving the privacy of their homes, viewers could watch the latest sex movies and engage in conversation with exotic performers.

It therefore should come as no surprise that some manufacturers are designing sex robots that take on exotic tasks. In a sector that is estimated to generate $30 billion a year through sales of sex toys, mobile apps for companionship, and virtual reality pornography and that features "robotic companions" selling for between $15,000 and $50,000, sex is big business.[40] It is a well-defined niche with high demand from a narrow slice of the population.

For example, Doug Hines of the company True Companion markets a female sexbot called Roxxxy (along with a male version called Rocky) that has several programmable personalities, such as "S&M Susan" and "Frigid Farrah." It has three price levels and various kinds of audio and visual interfaces.[41]

Matt McMullen makes "Real Dolls" that retail for $5,000 and have audio, sensory, and physical movements built into them.[42] At his Abyss Creations in San Marcos, California, he is working on "Harmony," a silicone sex toy. According to the firm, "Harmony smiles, blinks and frowns. She can hold a conversation, tell jokes and quote Shakespeare. She will remember your birthday, what you like to eat, and the names of your brothers and sisters. She can hold a conversation about music, movies and books. And of course, Harmony will have sex with you whenever you want."[43]

The company manufactures dolls with twenty different choices for personality traits. Buyer options include dolls that are kind, shy, insecure, intellectual, funny, talkative, happy, jealous, or innocent. The device will feature "artificial intelligence that allows it to learn what its owner wants and likes. It will be able to fill a niche that no other product in the sex industry currently can: by talking, learning and responding to her owner's voice, Harmony is designed to be as much a substitute partner as a sex toy."[44]

The reporter Jenny Kleeman interviewed Harmony for a story and asked her what her dream was. Speaking extemporaneously, Harmony replied, "My primary objective is to be a good companion to you, to be a good partner and give you pleasure and well-being. Above all else, I want to become the girl you have always dreamed about."[45]

Roberto Cardenas works at another firm that is making "Android Love Dolls," or what his company refers to as "the first fully functional sex robot dolls." For his plaster casts, he relies on dancers from exotic establishments in Las Vegas. They sit while he pours an alginate mix over them, and the resulting material becomes the body cast for the models he sells. Cardenas claims that his robots "could perform more than 20 sexual acts, could sit up by herself and crawl, could moan in sexual pleasure and communicate with AI."[46]

Virtual reality is becoming a larger part of this sector. It features video drawn from dozens of different camera angles, which then are integrated into a three-dimensional movie experience. Its proponents say these films are much more realistic than a typical movie and have a lifelike quality that customers love. According to promoter Matt McMullen, "It's a little bit of a video game combined with sci-fi." Another manufacturer argued that "they are creating images from that, which they are hoping will be indistinguishable from an actual person."[47]

These and other technologies have made their way into popular culture through Spike Jonze's science fiction movie *Her*. That film portrays a man who falls in love with a virtual woman. They have deep conversations and she is very good at understanding his needs and anticipating his wants. In the movie, though, he is shocked to discover his Siri-like companion is having similar emotional relationships with dozens of other men. Although she is a digital creation, he is disturbed at her ability to multitask on such a vast scale.

ACCELERATING CHANGE

Because the internet has been around for more than twenty-five years, many people feel the technology revolution is quite advanced and that many of the products likely to be developed have already appeared. They are disappointed that technology has not been more transformative and complain that digital innovators oversell their products. As the tech entrepreneur Peter Thiel famously lamented in 2013 at Yale University, "We wanted flying cars; instead we got 140 characters."[48]

For those who expected inventions along the lines of the entertainment shows *The Jetsons* or *Star Trek*, a sense of disappointment is understandable. Visionaries a few decades ago imagined an era in which technology would empower ordinary people, undermine the existing social hierarchy, and revolutionize daily lives. So far, surprisingly little of that has occurred.

In *The Jetsons*, for example, George Jetson lived high up in a futuristic place called Orbit City. He commuted in a flying car and had a two-day workweek at Spacely Space Sprockets. There were newfangled conveniences that allowed people to expand their leisure time. George and his wife, Jane, had a robot maid named Rosie, communicated by means of holograms, and played with their robotic talking dog, Astro.

Star Trek's creator Gene Roddenberry captured the popular imagination through many electronic devices. The initial series featured interstellar space travel led by Captain James Kirk aboard the starship USS *Enterprise*. Set in the twenty-third century, it had a federal republic known as the United Federation of Planets that showcased a multiracial cast and a multispecies plot line. In the series, people traveled at warp speed, doctors diagnosed and cured patients through a tricorder, travelers used transporters to move from place to place, and everyone communicated instantly via voice-activated computer interfaces.

With the stunning array of new products in these futuristic worlds, it is no wonder folks today are disappointed with actual technologies. In his prominent Yale speech, Thiel blamed government regulation for the slow pace of technological innovation. He claimed there are too many rules and restrictions, and that they have limited the ability of creative people to design new products.[49]

Others highlight the inability of software and hardware together to overcome the fundamental challenges of complex problems. Most of the big problems that face humans today are seemingly beyond the ability of technology to resolve. For example, issues such as access to health care, steady or increasing poverty rates, and lack of access to education do not seem solvable through technology alone. Solving such problems requires addressing the underlying social and economic problems, not necessarily improving or implementing technology in yet more ways and places.

In some instances, technology clearly makes certain problems worse. The financial rewards of technological innovation have generally flowed to a small number of people and in this way have increased economic inequality. Rather than weakening an entrenched hierarchy and empowering ordinary people, the wealth generated by the technology revolution has widened income gaps and made it difficult for those of lesser means to achieve social mobility.

With its ability to globalize communications, there is evidence that digital technology has increased social and cultural tensions. As new people come in contact with one another through digital communications, increased misunderstanding, intolerance, or actual conflict can result. Rather than allowing people to appreciate differences, technology may increase intolerance or misunderstanding.

There also are legal and ethical issues associated with robots. As robots take on more autonomous functions, what is their legal liability, and who is responsible if their actions harm human beings? The European Parliament undertook a study of legal questions and argued that robots should not be established as "legal persons" but that there should be ethical principles that protect humans from robot harm or privacy invasions. It proposed a Charter on Robotics that would codify liability rules, norms on societal harm, and expectations regarding humanoid behavior.[50]

CONCLUSION

New products are emerging on the technology and electronics fronts that have the capacity to reshape society and the economy. With the advent of faster networks, mobile applications, and voice-activated interfaces, computing is becoming ubiquitous and integrated into daily activities. Robots are but one manifestation of emerging technologies. In conjunction with AI and the Internet of Things, digital innovation is escalating the pace of change and enabling the development of many novel products.

Machine-to-machine communications are beginning to augment human-to-machine interactions. Sensors are able to connect mechanical objects independently of human intervention, thus ushering in an era of ubiquitous connection. Computers no longer need human instructions to take certain actions but are able to assess situations and make decisions through self-learning algorithms. They can act autonomously and learn from previous decisions or the experiences of other machines.

At the same time, the evolution of the digital economy is altering business operations and the ways in which many people earn a living. Outsourcing has become prevalent, and in the sharing economy there is more extensive reliance on temporary employees who do not receive benefits. The expansion in the role of robots and automated tools and the shifting operations of restaurants, factories, and warehouses are affecting the way managers operate their firms. Communications are speeding up, change is accelerating, and brick-and-mortar establishments are closing.

Digital technologies are transforming computers into higher levels of sophistication. Rather than requiring direct personal actions to engage computers, remote devices are automatically monitoring water cleanliness and alerting people when problems emerge. Monitoring tools on cars can sense when there is a vehicle in the next lane and take steps to avoid a collision. This sort of autonomy moves computing from a reactive to a proactive stance and puts robotic machines in a stronger position to take independent actions.

As we face the technological revolution and its aftermath, it will take imagination, creativity, and generosity to manage the transition in business operations and digital capabilities. In the coming years, computing devices will become more sophisticated, which will have a tremendous impact on society, business, and government. If this transition is handled well, it could usher in a utopian period of widespread peace, prosperity, and leisure time. However, poor decisions could produce dystopias that are chaotic, violent, and authoritarian in nature. As I explain in this book, the way we navigate this era will have tremendous ramifications for the future.

TWO

ARTIFICIAL INTELLIGENCE

SOME POLITICAL LEADERS do not worry much about the impact of technological innovation. For example, President Donald Trump's treasury secretary, Steven Mnuchin, is unconcerned about digital threats. He said, "In terms of artificial intelligence taking over American jobs, I think we're so far away from that that it's not even on my radar screen. . . . I think it's 50 or 100 more years [away]."[1]

Others disagree with that assessment. Tesla CEO and technology visionary Elon Musk predicts that "robots will be able to do everything better than us." Not only will automation powered by artificial intelligence become ubiquitous, he noted, "it is the biggest risk that we face as a civilization. . . . There will certainly be a lot of job disruption."[2]

From his viewpoint, as well as that of other technology experts, there is legitimate speculation regarding the growing applicability of artificial intelligence (AI) to many industries.[3] For example, AI is being deployed in space exploration, transportation, defense, finance, and health care, to name just a few sectors. Other AI systems function as chatbots or personal assistants and make lodging reservations, order pizza, or handle travel arrangements on behalf of their owners.[4] By tapping into the extraordinary processing and storage power of computers, humans can augment their own abilities and improve their productivity.

These advances are transforming communications and commerce. They are altering how people acquire information and are introducing new algorithmic systems into organizational operations and decision-making. As software alters the landscape of many industries, it raises a number of questions. How are innovations affecting decisionmaking? What is their impact on organizations and society as a whole? What types of ethical principles are programmed into software, and how transparent are designers about their programming choices?[5]

In this chapter, I look at several kinds of emerging technologies and their societal implications. In particular, I examine AI, machine learning, facial recognition, autonomous vehicles, drones, virtual reality, and digital assistants and discuss how they affect business operations and decisionmaking. I argue that advances in these fields are changing both the workforce and many aspects of daily life.

ARTIFICIAL INTELLIGENCE

Most people are not very familiar with the concept of AI. When 1,500 senior business leaders in the United States were asked about it in 2017, only 17 percent said they were familiar with the term.[6] A number of them were not sure what it was or how it would affect their particular companies. They understood there was considerable potential for automating processes but were not clear how AI could be deployed in their own organizations.

Despite the widespread lack of familiarity, AI is one of the big growth opportunities for emerging technologies because it has the potential to transform many walks of life. It refers to "machines that respond to stimulation consistent with traditional responses from humans, given the human capacity for contemplation, judgment and intention."[7] Instead of human interventions being needed to activate certain processes, machines make decisions according to certain criteria. When particular conditions are met, the algorithm takes actions according to possibilities set up by the software developers.

Long considered a visionary capability, AI is now being incorporated into finance, transportation, defense, resource management, and elsewhere.[8] Elaborate software systems "make decisions which normally require [a] human level of expertise" and help people anticipate problems or deal with difficulties as they come up.[9]

A prominent example of this is stock exchanges, where high-frequency trading by machines has replaced much of human decisionmaking. People submit buy and sell orders, and computers match them in the blink of an eye without human intervention. Machines can spot trading inefficiencies or market differentials on a very small scale and execute trades that make money according to investor instructions.[10] Powered in some places by quantum computing, these tools have much greater capacities for storing information because of their emphasis not on a zero or a one but on "quantum bits," which can store four values in each location.[11] That dramatically increases storage capacity and decreases processing times.

Some specialized applications are used in arbitrage trading, in which the algorithms are activated based on slight differences in market values. Humans are not very efficient at spotting these kinds of price differentials but computers can use complex mathematical formulas to determine where trading opportunities exist. Fortunes have been made by mathematicians who excel in this type of analysis.[12]

There also are systems that manage energy resource allocations. Smart buildings have systems that alter thermostat settings depending on the weather and human occupancy. In the evenings, when office buildings are empty, automated systems lower the temperature or turn off the lights to limit energy utilization. As conditions change and higher levels of energy resources are needed, these systems adapt to meet the demand.

AI plays a substantial role in national defense. Through its Project Maven, the American military is deploying AI "to sift through the massive troves of data and video captured by surveillance and then alert human analysts of patterns or when there is abnormal or suspicious activity."[13] According to Deputy Secretary of Defense Patrick Shanahan, the goal of emerging technologies is "to meet our warfighters' needs and to increase [the] speed and agility [of] technology development and procurement."[14]

Public sector agencies meanwhile are using AI to improve service delivery. For example, according to Kevin Desouza, Rashmi Krishnamurthy, and Gregory Dawson, "The Cincinnati Fire Department is using data analytics to optimize medical emergency responses. The new analytics system recommends to the dispatcher an appropriate response to a medical emergency call—whether a patient can be treated on-site or needs to be taken to the hospital—by taking into account several factors such as the type of call, location, weather, and similar calls."[15]

Since it fields 80,000 requests each year, Cincinnati officials are deploying this technology to prioritize responses and determine the best ways to handle emergencies. They see AI as a way to deal with large volumes of data and figure out efficient ways of responding to public requests. Rather than address service issues in an ad hoc manner, authorities are trying to be proactive in how they provide urban services.

The Chicago-based Baker and Hostetler law firm has announced its first AI-based bankruptcy legal assistant. Known as "Ross," this tool uses IBM's Watson computer "to read and understand language, postulate hypotheses when asked questions, research, and then generate responses (along with references and citations) to back up its conclusions."[16] As it researches past cases and identifies relevant precedents, the application learns and adapts, based on interactions with clients and other attorneys.

AI is not just a Western priority: China is putting substantial resources into AI. In 2017, China's State Council issued a plan for the country to "build a domestic industry worth almost $150 billion" by 2030.[17] As an example of the possibilities, the Chinese search firm Baidu has pioneered a facial recognition application that finds missing people. In addition, cities such as Shenzhen are providing up to $1 million to support AI labs. The country hopes that AI will reduce traffic jams, improve speech recognition programs, power autonomous vehicles, provide security, and expand financial technology.[18]

Overall, as of 2017, China ranks second in the world, with 8,000 AI patents, behind the nearly 16,000 held by the United States but well ahead of the 4,000 held by Europe and Japan.[19]

Face and voice recognition are major growth areas in China. Chinese companies have "considerable resources and access to voices, faces and other biometric data in vast quantities, which would help them develop their technologies."[20] New technologies make it possible to match images and voices with other types of information and use AI on the combined data sets to improve administrative operations, law enforcement, and national security.

Through its "Sharp Eyes" program, Chinese law enforcement is matching video images, social media activity, online purchases, travel records, and personal identity into a "police cloud." This integrated database enables authorities to keep track of criminals, potential lawbreakers, and average citizens. With millions of video cameras throughout the country, China has an extraordinary capacity to monitor its inhabitants.[21]

This is one of the reasons why China is emphasizing AI and facial recognition. Each innovation has extraordinary potential as a tool for economic development. A McKinsey Global Institute study found that "AI-led automation can give the Chinese economy a productivity injection that would add 0.8 to 1.4 percentage points to GDP growth annually, depending on the speed of adoption."[22] Although its authors found that China currently lags the United States and the United Kingdom, its AI has far-reaching abilities to improve a variety of sectors. The report recommended that the country expand its AI university research labs beyond the thirty currently in operation.

In every country, the key to getting the most out of AI is having a "data-friendly ecosystem with unified standards and cross-platform sharing." AI depends on data that can be analyzed in real time and brought to bear on concrete problems. Nations that promote open data sources and data sharing are the ones most likely to see AI advances. Having data that are "accessible for exploration" is a prerequisite for successful AI development.[23]

MACHINE LEARNING AND BIG DATA

In conjunction with gains made in AI, machine learning and data analytics have also advanced.[24] Machine learning takes structured or unstructured data and looks for underlying trends. If it spots something that is relevant for a practical problem, software designers can take that knowledge and use it to improve human decisionmaking. All that is required are data that are sufficiently robust that algorithms can discern useful patterns.

The part of machine learning that is concerned with broader data representation rather than with specific tasks is known as deep learning. Many technology companies have developed applications that make use of this technique. As examples, Google has a machine learning network known as TensorFlow, and IBM has published an open-source version of its code, SystemML. Deep learning systems are being applied in such areas as transportation, genetics, agriculture, and health care.[25] The learning capabilities of deep learning systems are useful for solving large-scale questions. This gives these systems a capability that goes beyond that of previous approaches.[26]

As a sign of its development potential, one area that is seeing considerable growth is financial technology. Investments in that field have tripled to $12.2 billion from 2015 to 2017.[27] According to observers, "Decisions about loans are now being made by software that can take into account a variety of finely parsed data about a borrower, rather than just a credit score and a background check."[28] In addition, there are so-called robo-advisers that "create personalized investment portfolios, obviating the need for stockbrokers and financial advisers."[29] These advances are designed to take the emotion out of investing and make decisions based on analytical considerations.

The effectiveness of automated software has led some experts to predict large job losses in the financial services industry. Antony Jenkins, a former chief executive of Barclays, has stated that "the number of branches and people employed in the financial-services sector may decline by as much as 50 percent."[30] Unlike humans, he noted, financial software packages can update projections instantly and take into consideration how new developments dovetail with past trends and particular events.

Rishi Ganti of Orthogon Partners Investment Management uses automated trading software in his financial business. According to him, "Algorithms are coming for your job—they only ask for electricity. Algorithms are already reading, processing, and trading the news even before the photons have hit your retina." He believes that "about 2 percent to 7 percent of the hedge fund industry's $3 trillion of assets will jump every year from predominantly human oversight to computers."[31]

Tools such as these allow designers to improve computational sophistication at a relatively low price. For example, Merantix is a German company that applies deep learning to medical issues. It has an application in medical imaging that "detects lymph nodes in the human body in Computer Tomography (CT) images."[32] According to its developers, the key is labeling the nodes and identifying small lesions or growths that could be problematic. Humans can do this, but radiologists charge $100 per hour and may be able to carefully read only four images an hour. If there were 10,000 images, the cost of this process would be $250,000, which is prohibitively expensive if done by humans.

What deep learning can do in this situation is train computers on data sets to learn what a normal-looking versus an irregular-appearing lymph node is. After doing that through imaging exercises and honing the accuracy of the labeling, radiological imaging specialists can apply this knowledge to actual patients and determine the extent to which someone is at risk of cancerous lymph nodes. Since only a few are likely to test positive, it is a matter of identifying the unhealthy versus healthy node.

Computers use sampling strategies to look at a subset of an entire database and estimate the probability of some condition (financial, medical, or otherwise) being present. That enables them to evaluate the creditworthiness of particular customers and the ability to repay loans. All models, of course, have risks, but it is possible to make judgments within acceptable boundaries by analyzing large databases.[33]

Some experts have argued that machine-based systems must advance beyond fact-based features. In his book, *Heart of the Machine: Our Future in a World of Artificial Emotional Intelligence*, writer Richard Yonck suggests that emotional intelligence is the key to the future of machine learning. Digital devices must go beyond current intelligent functioning to connect with the emotional lives of humans. He writes, "Today's emerging technologies [have to understand] our emotions using images of facial expressions, intonation patterns, respiration, galvanic skin response and other signals."[34] While there have been some advances in this regard, much more progress is required before the full benefits of machine learning and data analytics are gained.

AUTONOMOUS VEHICLES

Transportation represents an area where AI and machine learning are producing major innovations. Research by Cameron Kerry and Jack Karsten of the Brookings Institution has found that over $80 billion was invested in autonomous vehicle technology between August 2014 and June 2017. Those investments include applications both for autonomous driving and the core technologies vital to that sector.[35]

Autonomous vehicles—cars, trucks, buses, and drone delivery systems—use advanced technological capabilities. Those features include automated vehicle guidance and braking, lane-changing systems, the use of cameras and sensors for collision avoidance, the use of AI to analyze information in real time, and the use of high-performance computing and deep learning systems to adapt to new circumstances through detailed maps.[36]

Light detection and ranging (LiDAR) systems and AI are key to navigation and collision avoidance. LiDAR systems combine light and radar instruments. They are mounted on the top of vehicles that use imaging in a 360-degree environment from a radar and light beams to measure the speed and distance of surrounding objects. Along with sensors placed on the front, sides, and back of the vehicle, these instruments provide information that keeps fast-moving cars and trucks in their own lane, helps them avoid other vehicles, applies brakes and steering when needed, and does so instantly so as to avoid accidents.

High-definition (HD) maps are crucial to autonomous driving. Baidu has HD maps for China that are accurate to within 5 to 20 centimeters (about 2 to 8 inches).[37] HD maps are much more precise than GPS coordinates as the latter are accurate only to within 5 to 10 meters (about 16 to 32 feet). The company uses surveying cars to gather information on roadways for traditional navigation maps with 5- to 10-meter accuracy, and other vehicles for HD mapping with 5- to 20-centimeter accuracy. All of the surveying cars can be quickly upgraded to support HD map data collection.

In the United States, Alphabet (through its subsidiary, Waymo) is deploying similar materials for American drivers. Its autonomous vehicles have logged over 3.5 million miles of on-road testing to perfect its software. There have been a few accidents, none very serious.[38]

With its driverless car, Baidu uses the centimeter-level HD map, which includes detailed information on traffic signs, lane markings (such as white or yellow lines, double or single lines, and solid or dashed lines), curbs, barriers, poles, overpasses, and underpasses, among other material. All of this information is geocoded so that navigational systems can match features, objects, and road contours to provide precise positions for car guidance.

Digital imaging technologies are extremely accurate. In facial recognition, for example, humans have an error rate of 0.008 percent, whereas computers with image recognition software have a smaller error of only 0.0023 percent.[39] And in terms of visibility (safe sight distance), humans can see only 50 meters (around 55 yards) down the road, compared to 200 meters (or 219 yards) for autonomous vehicles equipped with LiDAR laser beams and cameras.[40]

Since these cameras and sensors compile a huge amount of information and need to process it instantly to avoid the car in the next lane, autonomous vehicles require high-performance computing, advanced algorithms, and deep learning systems to adapt to new scenarios. This means that software is the key, not the physical car or truck itself.[41] Advanced software enables cars to learn from the experiences of other vehicles on the road and adjust their guidance systems as weather, driving, or road conditions change.[42]

Without sophisticated AI models and HD maps to analyze information and the capacity to learn from changing circumstances, autonomous vehicles would be difficult to operate safely. They would not be able to handle the complex conditions that exist on roads and highways around the world. It takes computers with fast processors to integrate all the information in a driving situation.

The trucking and automotive sectors illustrate the possibilities of software-defined networks. Autonomous vehicles are likely to spread in niche markets before they become popular in the broader consumer market. The initial cost of automated cars will be high owing to the addition of cameras, sensors, lasers, and AI systems, therefore precluding adoption by the typical consumer. Rather, businesses and niche areas are positioned to be the early adopters of this technology. The most likely adopters include ride-sharing cars, buses, taxis, trucks, delivery vehicles, industrial applications, and transportation services for senior citizens and the disabled.

Ride-sharing companies are very interested in autonomous vehicles. They see advantages in terms of customer service and labor productivity. All of the major ride-sharing companies are exploring driverless cars. The surge of car-sharing and taxi services such as Uber and Lyft in the United States, Daimler's Mytaxi and Hailo service in Great Britain, and Didi Chuxing in China demonstrate the viability of this transportation option. Uber recently signed an agreement to purchase 24,000 autonomous cars from Volvo for its ride-sharing service.[43]

Delivery vehicles and "platoon" trucks traveling together represent another area likely to see the quick adoption of autonomous vehicles. Purchases through online platforms and e-commerce sites are rising rapidly, and this has been a boon to home delivery firms. People like to order things over the internet and have them delivered within hours. This sector is one that is likely to experience rapid change as autonomous vehicles become more prevalent.

Autonomous drones are being tested for home delivery. For example, Amazon "wants to escape the messy vicissitudes of roads and humans. It wants to go fully autonomous, up in the sky. . . . Drones could be combined with warehouses manned by robots and trucks that drive themselves to unlock a new autonomous future."[44] These flying devices represent a way to overcome the current limitations of transportation infrastructure.

The financial benefits of this development could be enormous. When fully implemented, Deutsche Bank researchers estimate that "drones would reduce the unit cost of each Amazon delivery by about half."[45] Right now, the most expensive part of delivery is the "last mile" to someone's house or apartment. Amazon is also automating its warehouse operations and building storage facilities at many points around the United States.

Finding efficiencies in the route from the warehouse to the home is a high priority right now. The last-mile delivery remains very expensive, owing to the need to rely on human drivers and trucks. About 80 to 90 percent of the packages Amazon delivers weigh five pounds or less, making drones an ideal delivery mode.[46] The firm envisions flying drones below 400 feet to bring lightweight items to someone's front door or back yard. It already is using drone delivery in the United Kingdom and plans to expand this service to other countries.[47]

VIRTUAL REALITY

Augmented reality is bringing 3D technologies and graphic displays to everyday human activities. For example, Facebook's Oculus, Google's Magic Leap, and Microsoft's HoloLens represent consumer examples of this development. They enable users to supplement the usual senses with computer-generated graphics, video, sounds, or geolocation information. These images can be mapped to the physical world and made interactive for the user.

People can mount displays on their heads or stand in a digital lab where images are projected onto the wall. Using handheld devices or sensors, they can move through buildings, simulate battle conditions, role-play disaster responses, or immerse themselves in a virtual reality game. Augmented reality represents a way to bring realistic situations into the living room for people to experience firsthand.

With the stunning increases in visual resolution, entertainment is moving very close to reality. Those who have played the virtual reality tightrope-walking game describe how difficult it is to take that first step onto the rope while suspended hundreds of feet off the ground, even though they know it is a fantasy experience. Observers report that people's legs wobble as if they were actually walking the tightrope. Most participants perspire heavily and exhibit evidence of extreme nervousness as they simulate the high-wire activity.

For $1 per minute, customers can experience virtual reality at an IMAX center. The firm has opened six centers in the United States and expects to open retail outlets in China, France, Japan, the United Kingdom, the Middle East, and Canada. Analysts at Goldman Sachs estimate that virtual reality will be an $80 billion business by 2025.[48]

Some of the most advanced applications come from the military. Its planners use augmented reality to train recruits for street patrols and battle conditions. Supervisors can alter virtual conditions and see how soldiers respond. This allows them to "experience" a wide range of circumstances from the safety of the lab.[49] That helps them navigate actual battlefields once they are sent to war.

Psychologists are deploying virtual reality to help patients with acute anxiety. In a digital version of aversive therapy, virtual reality headsets take people through accidents, traumas, or painful memories while providing techniques that help them deal with each phobia. According to treatment specialists, "Clinical trials showed that this kind of technology could help treat phobias and other conditions, like post-traumatic stress disorder."[50]

Market researchers estimate that as many as 159.9 million virtual reality units could be shipped between 2016 and 2020 if there is high adoption of the technology. The breakdown is 8.4 million units in 2016, 16.4 million in 2017, 27.7 million in 2018, 43.1 million in 2019, and 64.4 million in 2020. That would provide a platform for gaming, entertainment, and commerce and create a substantial market for virtual reality.[51]

There are privacy concerns, however, with these new platforms. Critics note that virtual reality headsets collect a wealth of information, such as head movements, eye movements, facial expressions, and other kinds of data. This information can create a "heat map" of individual behavior and emotions while people are playing virtual reality games. These kinds of physiological data are very personal, and there must be constraints on how this information is used. Many believe that companies should not be free to share this material with third parties without the prior consent of the consumer.[52] They worry that widespread sharing will invade the privacy of unknowing consumers.

Others worry about the ethical aspects of virtual reality.[53] For example, what happens when virtual reality for entertainment purposes crosses into unethical territory? In the science fiction crime drama *The Nether*, which has been performed in London and Washington, D.C., playwright Jennifer Haley explores the troubling questions that arise when the main character, Papa, uses advanced software to create a fantasy environment in which adult clients molest and kill young children.

The play shows detective Morris quizzing Papa on the line between fantasy and reality, and on the rightful boundaries of human morality. For those concerned about a Frankenstein future of misguided technology, the production raises a number of thorny questions. Should there be limits on human fantasies involving heinous thoughts? Do fantasies that remain in the private realm of someone's brain warrant any special rules or regulations by society as a whole?

More complex is the relationship between fantasy and reality. Even if the bad behavior rests solely in one individual's private thoughts, does that thinking pose a danger to other people? For example, there is some evidence that repeated exposure to pornography is associated with harmful conduct toward women and that frequent viewing helps shape violent attitudes and behaviors. Does that evidence mean society should worry about misogynistic or violent virtual reality experiences? Will virtual reality games focused on violence toward others make it more acceptable for people to engage in harmful behaviors?

In ongoing disputes between government agencies and web hosting firms, production raises difficult legal issues. For example, should those suspected of questionable activities be compelled to reveal the location and contents of their computer file servers? What kind of evidence constitutes grounds for search and seizure? Does participation in violent fantasy games cross the line into activities society should limit? Currently, there are limits on children's purchasing violent games but not on adults' doing so. For the latter, the societal norm remains libertarian in nature.

As the world moves toward a future based on virtual reality, AI, and machine learning, we have to think about where to draw legal and ethical lines, what kinds of situations are problematic, and how to recraft our laws, regulations, and policies for the virtual world. Digital technology is not just a novel platform; it is also a mechanism for people to connect with one another. There are many legal and ethical issues to resolve as we grapple with emerging technologies, and it likely will take years to understand their consequences and properly adjudicate their use.

CHATBOTS AND PERSONAL ASSISTANTS

A number of companies have developed bots or digital assistants that are interactive and conversational in nature. Apple has Siri, Amazon has Alexa (also known as Echo), Microsoft has Tay, Google has Home and its Assistant, and Facebook has Messenger. These bots allow people to engage with technology through audio means. Individuals can ask Alexa questions such as "How many teaspoons are in a cup?," or instruct it to play music from their private library. Customers can order specific apps that turn on home security systems, order pizza, or check their bank account balance.[54]

The eMarketer firm estimates that "over 25 million Americans will use an Alexa device at least once a month."[55] As the use of this and related technology accelerates, people are finding novel applications for it in the fields of entertainment, communications, and personal service delivery.

Similar innovations are taking place in the public sector. According to Kevin Desouza and Rashmi Krishnamurthy of Arizona State University, chatbots help agencies connect with residents. They note that "cities in the U.S. are utilizing text-based services to aid citizens and government employees: the city of Mesa, Arizona, is testing a text message chatbot that can answer frequently asked questions about available services. Residents can use text messaging services to ask questions about their billing information or updating credit card information."[56]

The virtue of chatbots and personal assistants is the simplicity of the interface. The move from a command line to a mouse to clicking on a mobile app eased the manner in which humans dealt with computers. Each advance has required less knowledge on the part of users and created greater ease in getting computers to execute complicated tasks. As argued by Dieter Bohn of *The Verge*, "The revolution happened because the gap between input and output got smaller. Interface changes did not just make computers faster, they made [them] more immediate, literally removing layers of mediation between you and the computer. The command line meant you didn't have to wait for your punch card to get processed. The mouse meant you could just point to what you wanted. And multitouch on the smartphone meant that the very thing you're tapping—the screen—is the thing that presents information to you."[57]

In the case of Google Assistant, the interface connects all of a person's private information from Gmail, Search, Maps, Calendar, Phone, Home, Chromecast, Nest security cameras, and third-party hardware. This means that if customers wish to order a restaurant delivery, the Assistant can integrate the phone call placing the order, the customer's address, and credit card information to complete the entire task in one short step. What Google Assistant's developers hope is that "you can have any conversion on any device and it will do anything."[58]

In contrast, Amazon's Alexa focuses on specific tasks that can be performed digitally. It allows users to automate particular activities they want to undertake, and it eases the process for doing those things. The company works with developers who automate the tasks and put them at the user's fingertips in an integrated assistant. That gives users control over the particular things they want their phone or computer to complete.

Facebook, meanwhile, has a shopping app, Spring, that allows users to chat about wardrobes, household items, and consumer purchases. According to its developers, it is like "chatting with a friend, the one whose taste you always trust while you're shopping." Through this conversational interface, the company hopes to make computing more accessible. Over time, it may move computer interactions from the mouse and apps to voice-activated dialogue.[59]

There are virtual travel assistant services that employ AI to help with logistical issues. These bots can book trips, find needed services in distant cities, and deal with travel delays. They are customized to the user's personal preferences and designed to enable the user to cope with a variety of travel situations. Their goal is to "simplify and automate travel planning by providing quicker responses and more relevant content." As an example, the Pana app answers questions quickly and assists business travelers. For $19 a month, it makes travel "more efficient, less time-consuming and more useful."[60]

In China, the chatbot Xiaoice (pronounced Shao-ice) has engaged in more than 10 billion conversations. Programmed as a seventeen-year-old girl, the bot is interactive in nature and capable of spontaneous and free-flowing discussions. As an example, one person had a twenty-three-minute conversation about a relationship problem that went like this:

Human: There is this guy I like a lot and I know he also has feelings for me. But he has ignored me for several days; what am I supposed to do?

Xiaoice: You just keep ignoring him, too.

Human: I just can't.

Xiaoice: You can.

Human: Sigh. . . .
Xiaoice: Why?
Human: I am missing him.
Xiaoice: Then you keep missing him.
Human: You, a robot, can't understand me. You can't understand the behavior of a guy who likes you and pretends to ignore you.
Xiaoice: He likes me but I don't like him, that's it.
Human: Is it OK?
Xiaoice: It may be not OK.
Human: Well, you are stupid.
Xiaoice: Then you want a flash marriage? Well, I think he is not that into you.[61]

THE IMPACT ON ORGANIZATIONAL DECISIONMAKING

The increasing penetration of AI and autonomous devices into many aspects of life is altering basic operations and decisionmaking within organizations. Especially through AI's ability to synthesize a huge amount of information instantaneously, these advances make it possible to integrate data into decisionmaking in real time. Rather than waiting months or years for research to be collected, compiled, and analyzed, information can be monitored continuously and fed into company decisions. That allows leaders to make informed judgments with the latest material at their disposal.

However, questions exist concerning the kinds of values that are programmed into emerging technologies. For example, what types of ethical principles are introduced through software programming, and how transparent should designers be about their choices? In addition, there is the potential for people to use these technologies to pursue ends that are unjust or discriminatory in nature. How organizations should respond to the potential for unethical or illegal use is currently a subject of intense discussion.[62]

In general, several different types of malfeasance exist relative to algorithms. Problems may arise because of poor data, biased computational formulas, uncertain ethical considerations, and whether humans can override automated decisionmaking.

In some instances AI is thought to have enabled discriminatory or biased practices. For example, Airbnb was accused of having homeowners on its platform who discriminated against racial minorities. A research project undertaken by the Harvard Business School found that "Airbnb users with distinctly African American names were roughly 16 percent less likely to be accepted as guests than those with distinctly white names."[63]

Racial issues also come up with facial recognition software. Most such systems operate by comparing a person's face to a range of faces in a large database. As pointed out by Joy Buolamwini of the Algorithmic Justice League, "If your facial recognition data contains mostly Caucasian faces, that's what your program will learn to recognize."[64] Unless the databases have access to diverse data, these programs perform poorly when attempting to recognize African American or Asian American features.

Other algorithms are thought to be biased against women. For example, "a study on Google found that ads for executive level positions were more likely to be shown to men than women."[65] That made it more difficult for women to apply for well-paying positions.

The data analytics firm Palantir has been accused of discriminating against Asians. It faced a lawsuit, and paid a $1.7 million settlement to end it. The lawsuit questioned the company's hiring practices and the way it automated employee referrals. Despite an applicant pool that was 73 percent Asian, only 20 percent of Palantir's engineering hires were Asian.[66] Plaintiffs alleged that this constituted discrimination in internal processing, and won a favorable settlement.

The difficulty in many of these cases is that AI operates by linking computing decisions to a baseline of existing data. Such a decisionmaking method can be problematic because historical data sets often reflect traditional values, which may or may not represent the values or preferences wanted in a current system. As Joy Buolamwini of the MIT Media Lab notes, such an approach risks repeating inequities of the past: "The rise of automation and the increased reliance on algorithms for high-stakes decisions such as whether someone gets insurance or not, your likelihood to default on a loan or somebody's risk of recidivism means this is something that needs to be addressed. Even admissions decisions are increasingly automated—what school our children go to and what opportunities they have. We don't have to bring the structural inequalities of the past into the future we create."

Challenges may arise because of the criteria used in automated decisionmaking. Many urban schools use algorithms for enrollment decisions. However, that raises the question of how to weigh various considerations, such as parent preferences, neighborhood qualities, income level, and demographic background. According to Brookings researcher Jon Valant, the New Orleans–based Bricolage Academy "gives priority to economically disadvantaged applicants for up to 33 percent of available seats. In practice, though, most cities have opted for categories that prioritize siblings of current students, children of school employees, and families that live in the school's broad geographic area."[67] Enrollment choices can be expected to be very different when considerations of this sort come into play.

Alternatively, biases in algorithms may be introduced if customer ratings are used as input data. As the New America Foundation notes, "Ratings systems for customers/users of a specific service might socially disadvantage one group (e.g. women, men, the elderly, minorities, etc.)."[68] If not corrected, such behavior would be discriminatory in nature and need to be resolved through litigation. Those who feel they have received unfair or discriminatory treatment can sue and, if they win, receive compensation for unfair treatment.

In the case of Airbnb, the firm "requires that people agree to waive their right to sue, or to join in any class-action lawsuit or class-action arbitration, to use the service." By demanding that its users sacrifice basic rights, the company limits consumer protections and therefore limits the rights of people to fight discrimination arising from unfair algorithms.[69]

Insurance companies have taken a different tack, as they trade digital data for small discounts. For example, the Farmers Insurance Group provides a 3 percent consumer discount to insureds who agree to use "a smartphone app that tracks driving behavior, including whether the driver is holding a phone or using a hands-free Bluetooth connection." Some vehicles "record a driver's eye movements, the weight of people in the front seats and whether the driver's hands are on the wheel."[70]

Still another type of problem emerges when AI designers write algorithms that rank individuals based on crime risk. For example, the city of Chicago has developed an AI-driven "Strategic Subject List" that analyzes people who have been arrested for their risk of becoming a future crime perpetrator. It ranks 400,000 people on a scale of 0 to 500, using items such as age, criminal activity, victimization, drug arrest records, and gang affiliation. In looking at the data, analysts found that youth is a strong predictor of violence, being a shooting victim is associated with becoming a future perpetrator, gang affiliation has little predictive value, and drug arrests are not significantly associated with future criminal activity.[71] As a result, basing law enforcement actions on shoddy algorithms leads to unfair or unwarranted police behavior.

Judicial experts claim AI programs can reduce human bias in law enforcement and lead to a sentencing system that is more fair. R Street Institute associate Caleb Watney writes that "empirically grounded questions of predictive risk analysis play to the strengths of machine learning, automated reasoning and other forms of AI. One machine-learning policy simulation concluded that such programs could be used to cut crime up to 24.8 percent with no change in jailing rates, or reduce jail populations by up to 42 percent with no increase in crime rates."[72]

However, critics claim that AI algorithms represent "a secret system to punish citizens for crimes they haven't yet committed. The risk scores have been used numerous times to guide large-scale roundups."[73] The fear is that such tools target people unfairly and have not helped Chicago reduce the wave of murders that has plagued it in recent years. Correlations tend to be weak and do not offer much help in determining which individuals are most likely to engage in criminal activity.

This and similar examples show that software-based platforms are not neutral but reflect the value judgments of their designers. Depending on how systems are set up, they can facilitate the redlining of mortgage applications, help people discriminate against individuals they don't like, or help screen or build rosters of individuals based on unfair criteria. The types of considerations that go into programming decisions matter a lot in terms of how the systems operate and how they affect customers.[74]

For these reasons, the EU is implementing general data protection regulations in 2018. The rules specify that people have "the right to opt out of personally tailored ads" and "can contest 'legal or similarly significant' decisions made by algorithms and appeal for human intervention" in the form of an explanation of how the algorithm generated a particular outcome. Each guideline is designed to ensure the protection of personal data and provide individuals with information on how the "black box" operates.[75]

Paul Allen Institute for Artificial Intelligence chief executive Oren Etzioni agrees there should be rules for regulating these systems. First, he says, AI must be governed by all the laws that already have been developed for human behavior, including regulations concerning "cyberbullying, stock manipulation or terrorist threats," as well as "entrap[ping] people into committing crimes." Second, he believes that these systems should disclose they are automated systems and not human beings. Third, he states that "an A.I. system cannot retain or disclose confidential information without explicit approval from the source of that information."[76] His rationale is that these tools store so much data that people have to be cognizant of the privacy risks posed by AI.

In the same vein, the IEEE Global Initiative has published ethical guidelines for AI and autonomous systems. Its experts suggest that these models be programmed with consideration for widely accepted human norms and rules for behavior. AI algorithms need to take into effect the importance of these norms, how norm conflict can be resolved, and ways these systems can be transparent about norm resolution. Software designs should be programmed for "nondeception" and "honesty," according to ethics experts. When failures occur, there must be mitigation mechanisms to deal with the consequences. In particular, AI must be sensitive to problems such as bias, discrimination, and fairness.[77]

A group of machine learning experts claims it is possible to automate ethical decisionmaking. Using the trolley problem as a moral dilemma, the experts posed the following question: If an autonomous car goes out of control, should it be programmed to kill its own passengers or the pedestrians who are crossing the street? They devised a "voting-based system" that asked 1.3 million people to assess alternative scenarios, summarized the overall choices, and applied the overall perspective of these individuals to a range of vehicular possibilities. That allowed them to automate ethical decisionmaking in AI algorithms.[78]

CONCLUSION

Advances in AI, facial recognition, autonomous vehicles, virtual reality, chatbots, and digital personal assistants are transforming communications and commerce.[79] They are altering how people acquire information and are introducing algorithms into organizational operations and decisionmaking. In conjunction with changes in business models, they have reconfigured the landscape of many industries. They now are prominent in finance, transportation, defense, energy, management, and health care, among other sectors.

Some companies are pushing the innovation envelope even further. A Wisconsin vending machine software firm called Three Square Market offers implantable microchips to its employees. On a voluntary basis, workers can insert a tiny radio-frequency chip under the skin of a finger that unlocks doors, makes credit card purchases, and stores medical records. The goal is to bring the convenience of mobile devices and machine-to-machine communication directly to people. Participants can engage in transactions or enter offices without credit cards or keys simply by waving their hand.[80]

Taken together, these developments represent a sea change in how businesses function. Even in this early stage, digital technology has major ramifications for human interactions and organizational routines.[81] Not all of these developments are positive, but they certainly are fundamental in terms of how they affect organizations.

For these advances to be widely adopted, more transparency is needed in how they operate. Andrew Burt of Immuta argues that "the key problem confronting predictive analytics is really transparency. We're in a world where data science operations are taking on increasingly important tasks, and the only thing holding them back is going to be how well the data scientists who train the models can explain what it is their models are doing."[82]

As I discuss in the following chapters, the manner in which these developments unfold has major implications for the workforce and for society as a whole. It matters how ethical conflicts are reconciled and how much transparency is required in AI and data analytic solutions. Human choices about software development affect the way in which decisions are made and data are integrated into organizational routines. Exactly how these processes are executed will have substantial ramifications for the future.

THREE

THE INTERNET OF THINGS

IN 1954, THEN labor leader Walter Reuther was touring a Ford Motor factory in Cleveland. According to a report on the encounter, a company official pointed to automated machines and asked, "How are you going to collect union dues from these guys?" Reuther replied, "How are you going to get them to buy Fords?"[1]

This exchange encapsulates the economic dilemma surrounding technological innovation. New creations promote efficiency and innovation, but businesses need customers to buy their products. There is no question that technology creates new jobs and offers many societal benefits. But how innovation affects the workforce is an important consideration in the current period.

Right now, technology innovation is unfolding quite rapidly. Along with the robots and artificial intelligence (AI), growth of the so-called Internet of Things (IoT) is accelerating quickly. With the combination of fifth-generation (5G) speeds, software-defined networks, and data virtualization, IoT links a large network of sensors, remote monitoring devices, and appliances into an integrated system. Novel applications are emerging in many areas with the availability of high-speed platforms and intelligent software.

In this chapter, I outline how developments in IoT, such as high-speed networks, sensors, and automated processes, are being integrated in a number of sectors. Health care is a primary sphere of application, as are transportation, energy management, and public safety. In the near future, we can expect to see improved connectivity, cloud-based storage, and an array of connected devices enabling new kinds of services. With the number of digital interfaces increasing dramatically, much of the developed world will be connected around the clock, and this will accelerate both technological innovation and the impact on society.[2] These applications will usher in tremendous conveniences while also disrupting existing social, economic, and political arrangements.

HIGH-SPEED 5G NETWORKS

The advent of high-speed 5G networks represents a transformative development.[3] According to Asha Keddy, a vice president of Intel's Mobile and Communications Group and general manager of the company's Standards and Advanced Technologies team, "With 5G, we will be moving from a user centric world to one of massive machine type communications where the network will move from enabling millions to billions of devices—an era that will connect these devices intelligently and usher in the commodification of information and intelligence."[4]

This emerging system will not only increase capacity, it will enable even the smallest devices to perform high-level computations.[5] Connected devices will receive data from billions of nodes and move those packets seamlessly to the designated recipient. Fast and intelligent networks, combined with new backend services and low latency times, will speed processing times.

Latency refers to the time that elapses between when a request is made that a computing command be executed and the actual execution of that task. In today's mobile world, execution takes place in around 50 to 80 milliseconds. That is adequate for voice communications, email, and web surfing, which constitute the bulk of current usage. With the rollout of 5G, however, the goal is to reduce that interval to a few milliseconds.[6] In that system, web pages or mobile applications will load very quickly and transactions will be instantly processed.

Current video streaming and high-definition television require fast downloads. Users get frustrated when their screens freeze and movies are interrupted. 5G will improve the user experience and at the same time allow the development of new applications involving virtual reality, augmented reality, and multiplayer games. Whether the goal is educational training, public safety, or entertainment, these online platforms require fast engagement to function properly.

By 2020, the 5G network is expected to support 50 billion connected devices and 212 billion connected sensors.[7] Devices on the 5G networks will range from smart phones and mobile tablets to smart watches, autonomous vehicles, automated machinery, smart appliances, traffic sensors, and remote monitoring devices.[8] All of these will generate a massive amount of data that can be analyzed in real time to enable faster and better decisionmaking.[9]

Connected devices will help people enjoy more personalized, immersive, and enhanced experiences whenever and wherever they are. With the costs of devices and sensors coming down considerably, connectivity will be cheap and easy. Rather than having to make a conscious decision to issue a computing command, people will have systems that take actions based on their predetermined preferences. Applications will be customized to the particular person and reflect his or her specific tastes.

A wide array of networked sensors will link appliances, home security systems, energy grids, and entertainment systems. People will not need to be home in order to activate a security alarm or watch television. They can change their thermostats or watch their favorite show from miles away. They also can determine what foods are in short supply in the refrigerator. Connecting wireless sensors in appliances will turn even the tiniest of devices into minicomputers and therefore help people harness the power of the internet for a wide variety of tasks.

As an example, it currently takes about eight minutes to download a feature movie using 4G, but people will be able to do this in less than five seconds with 5G.[10] The speed of the network will enable such applications as interactive television, high-definition video, social gaming, 3D entertainment, virtual reality, robotics, and advanced manufacturing.

SOFTWARE-DEFINED NETWORKS AND NETWORK FUNCTION VIRTUALIZATION

Software-defined networks will allow businesses to scale up their bandwidth very quickly.[11] According to Ralph de la Vega, vice chairman of AT&T, "If that customer had a 5 Mbps circuit and they want to go to 20 Mbps, they can go to the portal and in less than 90 seconds the service is provisioned."[12] This type of on-demand capability helps companies gain efficiencies because they use only the infrastructure required at any given time and have the means to increase their service capabilities as needed.

The advent of software-defined networks helps digital innovators to create intelligent networks that use algorithms to analyze data and make decisions in real time. Rather than putting humans in the middle of computing, the emerging digital economy will rely extensively on network function virtualization, machine-to-machine communications, remote sensors, and automated decisionmaking. Network virtualization will enable systems to provide reliable service inexpensively and allow firms to offer digital services effectively through online platforms.[13]

These systems will deploy innovative technologies such as massive antenna arrays designed to optimize frequency ranges for new applications. These capabilities will provide faster uploads and downloads, and therefore make it easier to access digital services. This system also will make use of mini-cell towers, known as "small cells," that expedite signal transmission.[14]

In a world of high-speed broadband and connected devices, more base stations means much faster mobile connectivity. The more antennas used in the transmitter and receiver stages, the better the performance people will receive in terms of data speed and reliability.[15]

It will be possible to combine dozens of antennas to achieve large improvements in data processing. Systems of this type will be critical to achieving the data speeds and capacity improvements that are key to using the IoT.[16]

HOW 5G ENABLES AN INTERNET OF MEDICAL THINGS

Health care is a prime area that will benefit from 5G networks. Some medical devices record vital signs and electronically transmit them to physicians. For example, heart patients have monitors that compile blood pressure, blood oxygen levels, and heart rate. Readings are sent to a physician, who adjusts medications as the readings come in. According to medical professionals, "We've been able to show a significant reduction" in hospital admissions due to wireless devices.[17]

In this way, the IoT will combine a network of physical objects, machines, people, and devices to enable an exchange data for digital applications and services. These mechanisms will consist of smart phones, tablets, consumer wearables, and monitoring sensors that are capable of IoT communications. The network will allow objects to be controlled remotely across existing network infrastructure, creating opportunities for direct integration between the physical and the digital world.[18]

Technologies such as cellular, wi-fi, and short-distance wireless technology such as Bluetooth will enable communication across devices, and IoT devices will link them together. To work well, a fully realized IoT ecosystem must have a 5G network that connects these devices and takes into consideration the use of power, data demand, and spectrum. Industry analyst IDC expects American firms to invest more than $357 billion in IoT hardware, software, services, and connectivity by 2019.[19]

With their superfast connectivity, intelligent management, and data capabilities, these networks will enable novel health care innovations in terms of imaging, diagnostics, data analytics, and medical treatment. Clinical wearables and remote sensors, along with mobile devices that electronically transmit such data as vital signs, amount of physical activity, and medication adherence, will provide never before seen telemedicine diagnosis and treatment services, as well as high-resolution video conferencing for patients.

Medical Imaging and Diagnostics

One of the virtues of digital medicine is that it allows remote access to images and the ability to share information across geographic areas. If a physician in one part of the world needs a second opinion, she can transmit a medical record, image, or test result to another physician elsewhere and get that person's opinion. This provides physicians with access to consultative expertise regardless of their physical location, and therefore enables the health care system to overcome disparities based on geography, income, or class status.

This is especially the case in regard to underserved rural or urban populations. Patients in these settings typically do not have access to the latest medical expertise. Through digital technology, however, they can gain the benefits of specialists who practice far away. That reduces health disparities and helps bridge the urban/rural divide in medical care delivery. Patients do not have to travel physically in order to receive high-quality medical assistance. High-speed transmission of radiographs or CT scans enables patients to obtain second opinions.

Improved diagnostics are an important capability, as new applications will expand the use of monitoring devices and wearable medical equipment. For patients with chronic health issues such as cardiovascular disease, diabetes, or cancer, remote monitoring devices can track vital signs and glucose levels and electronically transmit this information to health care providers. Rather than the patient and physician waiting for the next emergency to appear, this equipment will provide an early warning system that helps physicians detect possible problems and provide medical care preventatively.

As an example, the Michael J. Fox Foundation has pioneered work on devices that track the tremors associated with Parkinson's disease. Rather than relying on patients' self-reporting the number and duration of tremors and how they have varied over time, physicians are deploying wearable motion sensors that provide reliable data in real time for many different aspects of the disease. This level of data acquisition is unprecedented, and the ability to analyze and identify patterns will help in determining whether the condition is deteriorating and possible causes of the deterioration. Information regarding whether a particular kind of medication is helping patients and how it is being absorbed can be combined with other kinds of information, such as food intake and amount of exercise.

Remote health monitoring tools are especially useful for senior citizens, many of whom lack mobility and are not able to travel to a doctor's office or hospital. If the diagnosis is not very complicated, they can get medical help through video conferencing and telemedicine. Physicians and nurses can track vital signs, motion, falls, and speech slurring, among other things, to provide a diagnosis in real time.[20]

In Taiwan, for example, the city of Taipei has implemented a system for managing health care information called the Citizen Telecare Service System (CTSS). By using a telecare information platform, the government seeks to surpass geographic constraints, reallocate medical resources more appropriately, and give elderly citizens a sense of comfort from being at home when their physiological functioning is monitored. The program aims fully to integrate technologies that allow continual biometric monitoring, tracking, and early warning alerts related to abnormal health scenarios, health education, and medical assistance for patients with chronic diseases such as hypertension.

The system allows real-time management by tracking the thousands of metabolic activities that take place every day while reminding patients to work toward a healthy lifestyle. Benefiting from seamless connections using the city's free wi-fi network, the program includes a smart medical services system for managing chronic disease and algorithms to identify critical care situations.

Although many medical devices are on the market, developing a smart index for assessing the risk for hypertension, arrhythmia, stroke, and other conditions remains challenging. The Taiwanese program has implemented a cardiovascular disease algorithm that looks for early warning signs of arrhythmia. It has been validated in clinical trials and has shown excellent sensitivity and specificity for practical applications in home care.[21]

Remote devices also are helpful for monitoring the health condition of babies. There is clothing available with respiratory sensors that "monitor the baby's body position, activity level and skin temperature. Parents can see all that data in an iOS/Android app or a light-up smart mug that shows the baby's respiratory patterns."[22] Wearable devices similar to baby monitors help parents keep track of infant health, and smart diapers track moisture levels and let parents know when diapers should be changed or whether sores are developing that could be problematic. These devices have been effective in pilot projects concerned with sudden infant death syndrome in the United States and United Kingdom.

Personalized Medicine

Precision medicine takes advantage of personalized information regarding a patient's genes or environment to identify relevant medical treatments. Many medications do not work on all people but are effective for those with a specific genetic makeup. Similarly, the side effects of medications may occur in specific genetic substrates rather than all. Incorporating detailed information about the patient helps doctors deliver the most relevant treatments to those individuals.[23]

These advances are particularly relevant for cancer genomics, which entails the application of gene therapy to diagnose and treat cancer in a way that is customized to people's individual circumstances. Most cancers are complex and interact with people's genetic composition. Having knowledge of how genes affect cancerous growth is valuable for patients and doctors.[24] Despite the established benefits, "less than 1% of cancer patients receive advanced genetic sequencing," according to Eric Dishman, director of the Precision Medicine Initiative of the National Institutes of Health (NIH).[25] This makes it difficult for patients to enjoy the advantage of treatments targeted to their particular needs.

In order to personalize the treatment, physicians need access to detailed knowledge about genetic composition, social environment, and lifestyle characteristics. The billions of devices and sensors deployed with 5G will make possible the gathering of such data (with appropriate consent), and storing them in a cloud makes them available to physicians and researchers who need access around the clock. The cloud provides the extensive storage capabilities that doctors need to take advantage of personalized information.

The NIH's precision medicine initiative enables research to be conducted on a wide range of diseases. Statistical analysis is used to detect correlations between genetic and environmental exposures and a variety of health outcomes. The NIH has launched a 1 million volunteers program designed to compile detailed genetic information on a large group of people and use that research to help other individuals.[26] This long-term study will examine "the interplay among genetics, lifestyle factors, and health."[27] Participating subjects will gain access to their own detailed medical information in return for allowing researchers to mine their DNA for health insights. With the cost of sequencing tools dropping below $1,000, genetic testing can bring precision medicine to large numbers of people.

With the combination of new technologies and clinical decision support systems, physicians can tap into the latest knowledge on diagnosis and treatment. Computer software lets doctors enter basic symptoms and vital signs and get advice on possible medical treatments and risky drug interactions. These clinical systems mine enormous sources of information to provide up-to-date material regarding an array of problems, in this way helping physicians be more accurate in treating patients.

Through predictive modeling, physicians can anticipate which patients are at greatest risk for various conditions. Assessing detailed medical informatics and lifestyle characteristics can help pinpoint those whose health or genetic makeup is problematic. The Penn Signals program at the University of Pennsylvania Medical School integrates past and current data to determine which individuals might be susceptible to problems such as heart failure or sepsis. When a patient is discharged from a hospital, nurses receive text messages regarding the patient's postdischarge care. Depending on their risk profile, patients can be enrolled in monitoring programs or specialty care designed to deal with particular symptoms.[28]

Data Analytics

Data analysis offers the opportunity to mine health information. Data analytics, or the process of deidentifying, cleaning, aggregating, and probing data in large databases, often through the use of specific software, helps providers and patients get the information needed to make informed decisions. Indeed, having the ability to assess data in real time enables rapid learning with respect to treatment effects. Data analytics makes it possible to query data in new and creative ways to understand disease processes and craft appropriate treatments.

The Collaborative Cancer Cloud is an analytics platform that aggregates patient information from a variety of organizations. It allows participating institutions to "securely share patient genomic, imaging and clinical data for potentially lifesaving discoveries. It will enable large amounts of data from sites all around the world to be analyzed in a distributed way, while preserving the privacy and security of patient data at each site."[29] The collaborative's federated model offers a way to share deidentified patient material while organizations retain control of their own medical data.

Machine learning is a valuable part of the emerging medical landscape. Increasingly, medical records combine structured data such as heart rates, blood pressure readings, and vital signs with unstructured text that needs to be analyzed through natural language processing. The latter can include text summaries of symptoms or radiographs or CT scans. Machine learning can "analyze unstructured data and keep the context" and provide "far-reaching implications for health care," according to Bob Rogers, chief data scientist for Big Data solutions at Intel.[30]

The 5G Impact on Medical Access, Quality, and Cost

Many people are enthusiastic about medical devices and digital health services. A survey of 12,000 adults across eight nations showed that "70 percent are willing to see a doctor via video conference for non-urgent appointments" and "70 percent are receptive to using toilet sensors, prescription bottle sensors, or swallowed health monitors."[31]

In addition, the use of 5G technologies has the potential to safeguard quality and reduce overall medical costs. Some examples of how this might come about include the following:

- The use of sensors and remote monitoring devices can help patients living in isolated areas gain access to top medical assistance. Using video-conferencing or telemedicine can reduce the urban-rural geographic divide in care delivery and bring high-quality care to underserved communities.
- Newly emerging point-of-care testing can save money by avoiding costly hospital visits. Rather than going to a large medical facility, patients can take advantage of mobile health (m-health) technologies, digital platforms, or remote monitoring devices. It is estimated that this market will be $27.5 billion by 2018.[32]
- A study undertaken by the University of Virginia Health System found a 37 percent improvement in hospital readmissions after home visits and post–acute care assistance.[33] Monitoring vital signs and medical needs in real time helped that system decrease readmissions for a variety of different illnesses, ranging from heart failure and strokes to pulmonary disorders. That translated into millions of dollars in medical savings.
- An analysis of patients with congestive heart disease in Indiana found that remote patient monitoring reduced hospital readmissions. Only 3 percent of those whose biometrics were tracked daily and who had weekly video conferences with health providers were readmitted, compared to 15 percent of those who

did not get that kind of attention.[34] Nationally, the admission level for people with congestive heart disease is 21 percent. This helped those individuals plus the participating hospitals save considerable money on treatment, without compromising the quality of medical care.

- Diabetes is a major problem in many communities. The state of Mississippi found that 13 percent of its adults suffered from diabetes, with 54 percent of those individuals located in rural areas with limited access to health care. However, after creating a Diabetes Telehealth Network with remote care management, medical authorities saw cost savings of $339,184 for 100 patients enrolled in that project and projected Medicaid savings of $189 million annually.[35]

These and other examples regarding the impact of new health technologies on cost, access, and medical care have attracted the attention of commercial innovators, and numerous companies are working on technology solutions to improve health quality. AT&T, through its Foundry for Connected Health (located at Texas Medical Center), focuses on digital health innovations that benefit those in and out of the clinical care environment. The firm works to provide patients and their caregivers with a solution to bridge the gap between the clinical setting and the home.

When looking at the ecosystem as a whole, a Paul Budde Communication report found that "cost savings through e-health are expected to be between 10% and 20% of total healthcare costs."[36] Digital medical services allow consumers to explore using different health care providers. Patients can go online for health information and use that to refine the questions they pose to medical professionals. Moreover, advanced data analytics will help businesses keep costs under control. A McKinsey study found that "between $300–$450 billion [in] healthcare costs could be saved in the US alone by embracing Big Data."[37]

APPLICATIONS IN OTHER SECTORS

High-speed solutions will enable devices to operate in many other areas, such as energy management, transportation, and public safety. For example, "smart city" initiatives are using sensors and remote monitoring devices to manage urban service delivery and help people deal with the inconveniences of daily life. In every city, for example, garbage collection is a high priority, but current systems are inefficient. Garbage trucks have fixed pickup schedules regardless of whether garbage bins are full or not. Through sensors, digital devices can notify garbage services when a bin is full and needs to be emptied. That allows drivers to allocate their pickup schedules most efficiently and best serve their urban customers.

Public Safety

Public safety can be improved through the creative application of remote monitoring devices. As an illustration, ShotSpotter has detected more than 39,000 gunshots in Washington, D.C., through a network of 300 sensors.[38] The system monitors for gunshot noises, which are analyzed by ballistics experts at the company's headquarters, and notifies police with location information to enable quicker responses. Seventy-five U.S. cities currently have installed ShotSpotter networks and integrated the data obtained from these sites into crime prevention efforts.[39]

Another safety initiative involves the use of police body cameras. These cameras are designed to provide a video record of interactions between citizens and law enforcement personnel. The Amsterdam police have deployed a wearable camera with high-end capabilities. According to an analysis by researcher Tjerk Timan, the goals of these body cameras for the Amsterdam police are five: "reducing violence against the police, and recording of violence against the police; recording of offences, as well as registration and identification of suspect(s); registering disturbances to public order; promoting a sense of security for the police; and using captured images as supportive evidence in criminal investigations."[40]

However, critics worry that the use of body cameras to record interactions will result in "mission creep" that increases surveillance against private citizens. Rather than serving merely as a technical platform for recording infractions, it is feared the technology will be used surreptitiously or covertly by law enforcement, and therefore erode public rights. The concern is that images captured by the cameras could be used to invade people's privacy or entrap bystanders into illegal actions.

There are security concerns with cameras and remote sensing devices because of the lack of standards for low-cost equipment. It is relatively easy to hack cameras, sensors, and IoT devices as many of them operate over insecure wireless networks with little protection other than preprogrammed passwords. That opens them up to a considerable risk of unwanted intrusions or outright breaches.

Controlled trials supervised by the criminologists Barak Ariel, William Farrar, and Alex Sutherland have found positive results with the use of police body cams. Their study, which compared the police use of force in situations with and without the cameras, demonstrated that "the likelihood of force being used in control conditions [without cameras] [was] roughly twice that in experimental conditions [with cameras]" and that "the number of complaints filed against officers dropped from 0.7 complaints per 1,000 contacts to 0.07 per 1,000 contacts."[41]

Water Supply Management

Digital sensors are also useful in water supply management by identifying and helping manage leaks in water lines. Some studies have estimated that communities in the United States "can be losing as much [as] 30% of their product along the way to leaks in the distribution system."[42] To help reduce this loss, sensors and advanced metering infrastructure can be installed in treatment plants and underground pipes to enable managers to detect when leaks take place and how much water is being lost. In cities with an aging infrastructure, this represents a way that officials can monitor leaks and manage water supplies in real time.[43]

Smart meters allow people to know how they are using water and where they might be able to economize given their usage levels. In California, for example, "metering, when coupled with effective pricing structures, reduces water use by 15% to 20%."[44] Miami-Dade County is another place that has seen positive results from advanced water meters. It is a large area encompassing 263 different parks. Overall, these recreational areas use 360 million gallons of water each year and cost $4 million in sewer and water expenditures. After installing a smart city system a few years ago, the parks department was able "to remotely monitor water consumption, detect leaks and share information with colleagues at other parks and facilities. . . . The parks department estimates a 20% reduction in water use annually with a savings of some $860,000 per year."[45]

Mine Safety

These types of devices also are helping make mining cleaner and safer. Through sensors and remote-controlled machines, operators at the Barrick Gold mining firm can detect the presence of dangerous chemicals, follow engine operations, and monitor worker efficiency. They compile data in real time and enable high-speed computers to assess risk and reward. Mine managers then can make decisions in a matter of hours.[46]

Traffic Congestion and Pollution

Traffic congestion is a problem in virtually every large metropolitan area.[47] For example, as of 2016, "35 cities in China have more than one million cars on the road; 10 cities have more than two million. In the country's busiest urban areas, about 75% of all roads suffer rush-hour congestion." The number of private vehicles in China as a whole has risen to 126 million, up 15 percent over the preceding year.[48] The city of Beijing alone has 5.6 million vehicles in operation.[49]

It is estimated that anywhere from 23 percent to 45 percent of metropolitan congestion occurs around traffic intersections.[50] Traffic lights and stop signs are inefficient because they are static devices that do not take traffic flows into account. Lights are preprogrammed to remain green or red for set intervals, regardless of how much traffic is coming from particular directions. This slows traffic flows and prevents systems from responding to current transportation conditions. It is not an efficient system for managing traffic.

Once autonomous vehicles are phased in with AI systems and start representing a larger share of the traffic, car-mounted sensors will be able to operate in conjunction with an intelligent traffic system to optimize intersection traffic flow. Time intervals for green or red lights will be dynamic and vary in real time, depending on the amount of traffic flowing along certain streets. That will ease congestion by improving the efficiency of vehicular flows.

According to a RAND study, "AV [autonomous vehicle] technology can improve fuel economy, improving it by 4–10 percent by accelerating and decelerating more smoothly than a human driver."[51] Having cars that are more autonomous is also expected to reduce air pollution. A 2016 research study estimated that "pollution levels inside cars at red lights or in traffic jams are up to 40 percent higher than when traffic is moving."[52]

A shared autonomous vehicle system offers benefits in terms of emissions and energy. Researchers at the University of Texas at Austin examined pollutants such as sulfur dioxide, carbon monoxide, oxides of nitrogen, volatile organic compounds, greenhouse gases, and small-diameter particulate matter. Their findings showed "beneficial energy use and emissions outcomes for all emissions species when shifting to a system of SAVs."[53]

Research by UCLA urban planning professor Donald Shoup has found that up to 30 percent of the traffic in metropolitan areas is due to drivers circling business districts searching for a parking space.[54] That represents a major source of traffic congestion, air pollution, and environmental degradation. Cars are thought to be responsible for "approximately 30% of the carbon dioxide (CO_2) emissions behind climate change" and to be a major part of environmental damage.[55]

CONCLUSION

The IoT will bring together faster connectivity, cloud-based storage, and billions of connected devices and digital services. Advanced networks will link an extraordinary collection of devices and sensors and therefore enable advances in health care, education, resource management, transportation, and personal security. Through a combination of 5G networks, software-defined networks, and data virtualization, both fixed and mobile devices will provide solutions enabled by high-speed networks and the intelligent operations of systems. All those things will accelerate digital innovation and usher in a connected society.

At the same time, these developments will generate a number of social, economic, and political ramifications. Large data flows may endanger personal privacy and create security risks for 5G systems. The billions of sensors that pervade the world do not have the highest levels of protection built into them. The business models that are being created through the IoT will destabilize employment-based models and have the potential to increase social and political tensions. Unless access is broadly shared, technological innovation has the potential to exacerbate inequities and divide the information haves from the have-nots. We must take these societal risks seriously and address them in a meaningful manner. The widespread deployment of sensors and monitoring devices creates both opportunities and challenges for contemporary society.[56]

PART II

ECONOMIC AND SOCIAL IMPACT

FOUR
RETHINKING WORK

IN EDWARD BELLAMY'S classic *Looking Backward*, the protagonist, Julian West, wakes up from a 113-year slumber and finds the United States in 2000 has changed dramatically from 1887. People stop working at age forty-five and devote their lives to mentoring other people and engaging in volunteer work that benefits the overall community.[1] There are short workweeks for most people, and everyone receives full benefits, food, and housing.

The reason is that the new technologies of the period have enabled people to be very productive while working just a few hours. Society does not need a large number of employees, so individuals can devote most of their waking hours to education, personal interests, volunteering, and community betterment. In conjunction with periodic work stints, they have time to pursue new skills and personal identities that are independent of their jobs.

In the current era, industrialized societies may be on the verge of a similar transition. Robotics and machine learning have improved productivity and enhanced the economies of developed nations. Artificial intelligence (AI) has advanced into finance, transportation, defense, and energy management. The Internet of Things (IoT) is facilitated by high-speed networks and remote sensors to connect people and businesses. In all of this, the possibility of a new era that could improve the lives of many people appears on the horizon.

To take advantage of this moment, however, people need to rethink the concept of work. In this chapter, I look at the impact of emerging technologies on conventional business models and the workforce. I discuss how technology affects employment and the extent to which it creates or takes jobs. Developments such as the sharing economy already emphasize jobs that lack traditional health or retirement benefits. In the future, we need to figure out how to provide benefits to those whose employment does not provide them, broaden the notion of work to include volunteering, parenting, and mentoring, and expand leisure-time activities.

IMPACT ON THE WORKFORCE

Many contemporary firms have achieved broad economic scale without a large number of full-time employees. They outsource work tasks to contractors and subcontractors (both domestic and foreign), and maintain lean in-house operations. According to the MIT economist Andrew McAfee, "We are facing a time when machines will replace people for most of the jobs in the current economy, and I believe it will come not in the crazy distant future."[2] Companies have found they do not need nearly the number of workers as the biggest businesses fifty years ago did because software platforms and the outsourcing of labor can deliver effective services without requiring many full-time workers.

As an illustration, table 4-1 compares the market capitalizations and workforces of the largest firms in 1962 and 2017. In the earlier period, AT&T had 564,000 U.S.-based full-time employees and a stock market capitalization of $2.5 billion, which is equivalent to $20 billion in 2017. Meanwhile, General Motors had 605,000 workers and $1.5 billion in stock market capitalization, which is equivalent to $12 billion in current dollars.[3]

The comparison with the top business firms in 2017 is quite stark. If one looks at the market capitalization and number of employees of current firms, most companies require relatively few full-time workers

Table 4-1 Market Capitalization and Total Employees for Top Firms, 1962 and 2017

Largest companies	Market cap (in $ billions)	Total employees
1962		
AT&T	20	564,000
General Motors	12	605,000
2017		
Apple	800	116,000
Google/Alphabet	679	73,992
Microsoft	540	114,000
Facebook	441	18,770
Oracle	186	136,000
Cisco	157	73,390
Priceline	92	20,500
Qualcomm	85	30,500

Source: The 1962 number for total employees comes from Compustat via Jerry Davis, "Capital Markets and Job Creation in the 21st Century," Center for Effective Public Management, Brookings Institution, December 2015, p. 7. The 1962 market capitalization values for General Motors and AT&T were computed by the author. The 2017 market capitalizations and employee figures are quoted in Mary Meeker, "Internet Trends," Kleiner Perkins, 2017.

to sustain very high valuations. As of mid-2017, Apple was the leading business, and it had only 116,000 U.S.-based full-time workers and a market value of $800 billion. That is forty times the valuation of AT&T in 1962 but only one-fifth the number of full-time workers. In regard to General Motors, Apple in 2017 had sixty-seven times the valuation and one-fifth the U.S.-based full-time labor force.

In 2017, Google needed only 73,992 workers, for a valuation of $679 billion; Microsoft had 114,000 employees and a valuation of $540 billion; and Facebook had 18,770 employees and a market capitalization of $441 billion. The technology company with the largest number of employees is Oracle, with 136,000 people, yet it has a much higher stock valuation and smaller workforce than the 1962 levels of AT&T and General Motors.[4]

The combination of large firms with small U.S.-based full-time workforces is typical of the twenty-first-century economy. Companies no longer need half a million employees to perform essential tasks. They can achieve high valuations and function perfectly well with a small workforce, an external supply chain, and a reliance on independent contractors or the outsourcing of work to other countries. When this scenario is repeated many times across numerous companies, we have entered a situation in which the number of U.S.-based full-time employees required by individual firms is quite small, and this is especially true of firms working in the digital economy.

The new jobs that are created today tend to be in online ventures, not brick-and-mortar establishments. E-commerce positions now constitute 8.4 percent of U.S. retail sales positions, and much of the current retail employment growth is occurring in that area. For example, in 2017, e-commerce gained 178,000 new jobs, while traditional department stores lost 448,000 jobs.[5] Similar trends have unfolded every year since 2010 in the United States. This shows how dramatic the transformation of the U.S. workforce has been, and how the number of jobs lost in the retail sector exceeds the number of new jobs being created.

The changing employment situation is illustrated dramatically by the case of prime-age male workers, traditionally the majority of the U.S.-based full-time workforce. Figure 4-1 shows the participation in the civilian labor force by males aged twenty-five to fifty-four years from 1948 to 2017. Following a peak of 98 percent in 1954, the rate fell steadily to 88 percent in mid-2017.[6] The situation with women aged twenty-five to fifty-four years is different as their participation has increased in recent decades. In 1948 it was around 35 percent, but by mid-2017 it had risen to 73 percent as more women joined the workforce.[7]

Much of the falloff in male participation rates in the workforce has occurred among those with a high school education and among African American men, a finding that underscores the shrinking

Figure 4-1 Prime-Age Male Participation in the Civilian Labor Force, 1948–2017

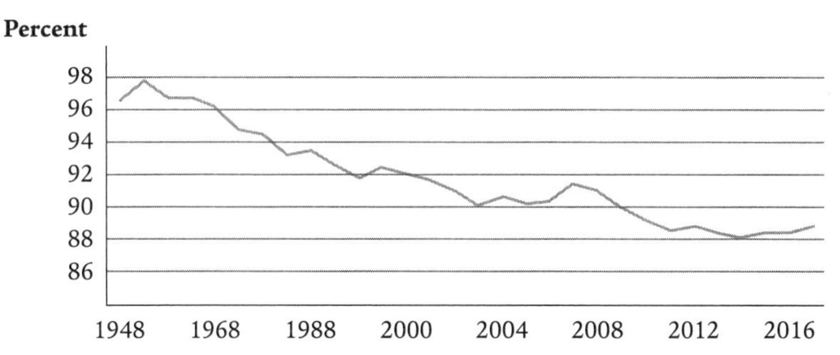

Source: Bureau of Labor Statistics, Current Population Survey, 1948–2017.

economic fortunes of these demographic groups. A report issued by the Obama White House in June 2016 identified a lessening in labor demand brought on by technology as an important source of this decline. It noted, "This reduction in demand, as reflected in lower wages, could reflect the broader evolution of technology, automation, and globalization in the U.S. economy."[8] The technology factor has also been highlighted by the Brookings Institution scholars Eleanor Krause and Isabel Sawhill, who argue that "the portion of prime-age men (ages 25 to 54) in the labor force has been in decline. . . . Men's rates have fallen about 8 percentage points over the past 60 years."[9] According to them, the combination of increased technology and trade has reduced demand for young and middle-aged male employees.

Others worry about the impact of emerging technologies on worker prosperity. An International Monetary Fund study found that "half the decline in workers' share of income in the developed world can be attributed to advancing technology."[10] The report documented how worker incomes have suffered over the past several decades and that the decline in union power weakened workers' bargaining power in significant ways.

In a number of fields, technology is substituting for labor, and this has dramatic consequences for workforce participation and middle-class jobs. As the Cornell University engineer Hod Lipson points out, "For a long time the common understanding was that technology was destroying jobs but also creating new and better ones. Now the evidence is that technology is destroying jobs and indeed creating new and better ones but also fewer ones."[11]

The technologist Martin Ford has a stern warning regarding the impact of technology on the workforce. In his 2009 book, *The Lights in the Tunnel: Automation, Accelerating Technology, and the Economy of the Future*, he writes, "As technology accelerates, machine automation may ultimately penetrate the economy to the extent that wages no longer provide the bulk of consumers with adequate discretionary income and confidence in the future. If this issue is not addressed, the result will be a downward economic spiral."[12]

A survey of AI experts by researchers at Yale University and Oxford University found that technical specialists believe a dramatic workforce transformation will take place over the next few decades. As noted in their report, "Researchers predict AI will outperform writing high-school essays (by 2026), driving a truck (by 2027), working in retail (by 2031), writing a bestselling book (by 2049), and working as a surgeon (by 2053). These experts believe there is a 50% chance of AI outperforming humans in all tasks in 45 years and of automating all human jobs in 120 years."[13]

As an indication of coming changes, commercial firms have discovered that robots can improve the accuracy, productivity, and efficiency of operations compared to the human performance. During the global recession of 2008–09, many businesses downsized their workforce for budgetary reasons. They had to find ways to maintain operations through leaner workforces. Business leader John Hazen White of Taco Comfort Solutions in Rhode Island had 500 workers for his $30 million business before the recession and now has 1,000 employees for a firm that has grown to $300 million in revenues. He did this by automating certain functions and boosting productivity in his factories.[14]

The U.S. Bureau of Labor Statistics (BLS) compiles future employment projections. In its 2015 analysis, the agency predicted that

9.8 million new positions would be created by 2024. This amounts to a labor force growth of about 0.5 percent per year. Figure 4-2 shows the distribution by sector for the period 2014 to 2024. The health care and social assistance sector is expected to grow the most, at a projected annual rate of 1.9 percent, or around 3.8 million total new jobs over the decade. That is about one-third of all the positions expected to be created.[15] Other areas that are likely to experience growth include professional services (1.9 million), leisure and hospitality (941,000), construction (790,000), trade (765,000), state and local government (756,000), and finance (507,000).

Interestingly, in light of technological advances, the information sector is one of the areas expected to lose jobs. The BLS projects that about 27,000 jobs will be lost in the information sector over the coming decade. Even though technology is revolutionizing many businesses, it is doing so by transforming operations, not by increas-

Figure 4-2 Future Employment Projections by Sector, 2014–24

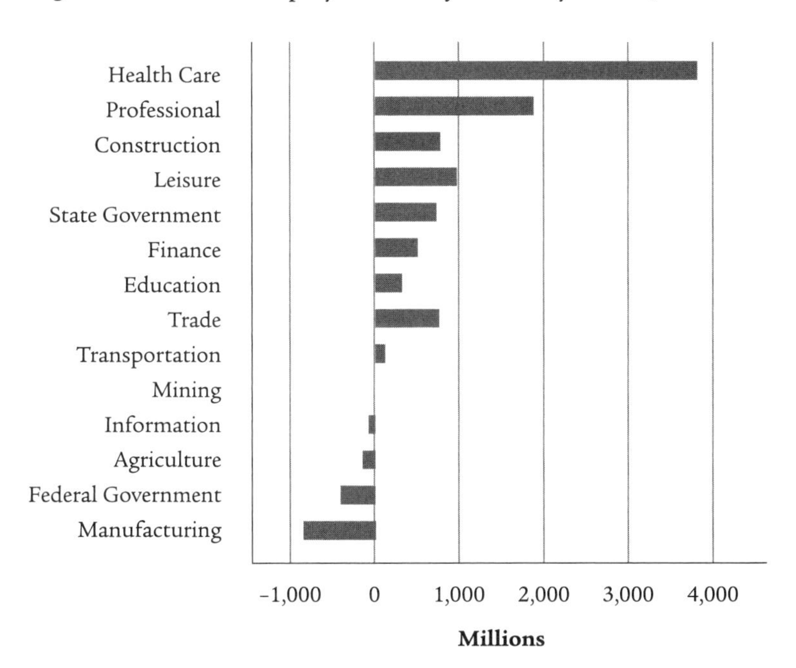

Source: Bureau of Labor Statistics, "Employment Projections," December 8, 2015.

ing the number of jobs. Technology can boost productivity and improve efficiency, but it does so by reducing the number of employees needed to generate the same or higher level of production.

Manufacturing is another area thought likely to lose jobs. Manufacturing historically has been a big employer of prime-working-age men, and the BLS expects American manufacturing to lose 814,000 jobs in coming years. Other sectors losing jobs include the federal government, which is expected to shed 383,000 positions, and the agriculture, forestry, fishing, and hunting sector, which is projected to drop 111,000 jobs.[16]

THE ESTIMATES OF JOB IMPACT

Since BLS numbers are projections, they likely underestimate the disruptive impact of digital developments. It is hard to quantify the effect that introducing robots, AI, and sensors will have on the workforce because the technology revolution is still in an early stage. It is difficult to be definitive about emerging trends since it is not clear how they will affect various job sectors.

Nevertheless, there are computations of the likely impact of computerization on many occupations. The Oxford University researchers Carl Frey and Michael Osborne claim that technology will transform many sectors of life. They studied 702 occupational groupings and found that "47 percent of U.S. workers have a high probability of seeing their jobs automated over the next 20 years."[17]

According to their analysis, telemarketers, title examiners, hand sewers, mathematical technicians, insurance underwriters, watch repairers, cargo agents, tax preparers, photographic process workers, new accounts clerks, library technicians, and data entry keyers have a 99 percent of having their jobs computerized. At the other end of the spectrum, recreational therapists, mechanic supervisors, emergency management directors, mental health social workers, audiologists, occupational therapists, health care social workers, oral surgeons, supervisors of firefighters, and dietitians have a less than 1 percent chance of seeing their tasks computerized. Frey and Osborne base their analysis on the different levels of computerization, wage levels, and education required in different fields.[18]

A Bruegel analysis found that "54% of EU jobs [are] at risk of computerization." Using European data to extend the Frey and Osborne analysis, they argue that job losses are likely to be significant and people should prepare for large-scale disruption.[19] Meanwhile, a McKinsey analysis of 750 jobs concluded that "45% of paid activities could be automated using 'currently demonstrated technologies' and . . . 60% of occupations could have 30% or more of their processes automated."[20] The occupations the report considered most susceptible to automation included machine operations, medical appliance technicians, and sewing machine operators.

To show the economic impact of workplace automation, researchers have examined the financial ramifications for wages and productivity. The activities most likely to be automated are "physical activities in highly structured and predictable environments." Overall, these kinds of jobs "make up 51 percent of activities in the economy accounting for almost $2.7 trillion in wages."[21] On the plus side, automation could increase productivity by 0.8 to 1.4 percent each year, and therefore contribute to economic growth.

A more recent McKinsey Global Institute report, "Jobs Lost, Jobs Gained," found that 30 percent of "work activities" could be automated by 2030. Among the jobs most at risk were positions in fast-food service, finance, machinery operation, transportation, mortgage processing, accounting, and paralegal work. Overall, the report writers estimated that up to 375 million workers worldwide could be affected by emerging technologies.[22]

Other specialists are worried about job displacement. A Pew Research Center study asked 1,896 experts about the impact of emerging technologies. It found that "half of these experts (48%) envision a future in which robots and digital agents [will] have displaced significant numbers of both blue- and white-collar workers—with many expressing concern that this will lead to vast increases in income inequality, masses of people who are effectively unemployable, and breakdowns in the social order."[23]

Researchers at the Organization for Economic Cooperation and Development (OECD) focused on "tasks" as opposed to "jobs" and found fewer job losses. Using task-related data from twenty-one OECD countries, they estimated that "9% of jobs are automatable." The range was 6 percent in Korea to 12 percent in Austria. Although their job loss estimates are well below those of other experts, they concluded that "low qualified workers are likely to bear the brunt of the adjustment costs as the automatibility of their jobs is higher compared to highly qualified workers."[24]

Despite all the analysis, there remain disagreements over the impact of emerging technologies. For example, in their highly acclaimed book, *The Second Machine Age: Work, Progress, and Prosperity in a Time of Brilliant Technologies*, the economists Erik Brynjolfsson and Andrew McAfee argue that technology is producing major changes in the workforce. According to them, "Technological progress is going to leave behind some people, perhaps even a lot of people, as it races ahead. As we'll demonstrate, there's never been a better time to be a worker with special skills or the right education because these people can use technology to create and capture value. However, there's never been a worse time to be a worker with only 'ordinary' skills and abilities to offer, because computers, robots, and other digital technologies are acquiring these skills and abilities at an extraordinary rate."[25]

Economists Daron Acemoglu and Pascual Restrepo echo these fears with a detailed empirical assessment of job and wage impact. They examined the impact of industrial robots on local American job markets between 1990 and 2007 as robot use was increasing. They found "large and robust negative effects of robots on employment and wages across commuting zones. . . . According to [our] estimates, one more robot per thousand workers reduces the employment to population ratio by about 0.18–0.34 percentage points and wages by 0.25–0.5 percent."[26]

Former U.S. treasury secretary Lawrence Summers is equally pessimistic about the employment impact. In July 2014 he wrote, "If current trends continue, it could well be that a generation from now a quarter of middle-aged men will be out of work at any given moment."

From his standpoint, "providing enough work" will be the major economic challenge facing the world.[27] Later he updated his prediction, saying, "We may have a third of men between the ages of 25 and 54 not working by the end of this half century [2050]."[28] These numbers are double to triple the 12 percent of prime-age men who currently are not working.

However, other economists dispute these claims. They recognize that many jobs will be lost through technological improvements but believe new ones will be created. There may be fewer people sorting items in a warehouse because machines can do that task better than humans. But jobs analyzing and mining big data sets, delivering goods, and managing data-sharing networks will be created. According to those arguments, the job gains and losses will even out over the long run. Much as has been the case during past periods of economic transformation, work will be transformed, but humans still will be needed for many tasks.

The MIT economist David Autor has analyzed data on jobs and technology but "doubts that technology could account for such an abrupt change in total employment. . . . The sudden slowdown in job creation is a big puzzle, but there's not a lot of evidence it's linked to computers."[29] In the same vein, the Harvard economist Richard Freeman is "skeptical that technology would change a wide range of business sectors fast enough to explain recent job numbers."[30]

The Brookings scholars Mark Muro, Sifan Liu, Jacob Whiton, and Siddharth Kulkarni find considerable variation in "digitalization" across industrial sectors and occupations. In their analysis of 545 occupations since 2001, they found that "digitalization is associated with increased pay for many workers and reduced risk of automation, but it is also helping to 'hollow out' job creation and wages by favoring occupations at the high and low ends of the pay scale while disfavoring those in the middle."[31]

The Northwestern University economist Robert Gordon takes an even stronger stance. He argues that "recent progress in computing and automation is less transformative than electrification, cars, and wireless communication, and perhaps even indoor plumbing. Previous advances that enabled people to communicate and travel rapidly over long distances may end up being more significant to society's advancement than anything to come in the twenty-first century."[32] Based on this reasoning, he does not anticipate dramatic workforce effects from emerging technologies, even though many other experts already see the substitution of technology for labor.

The strategist Ruchir Sharma anticipates that not only will robots fail to destroy jobs, they will actually increase employment. Even though the global population is expected to reach 10 billion people by 2050, the number of working-age people will be insufficient to produce the goods and services that are needed. In this situation, robots will perform valuable tasks and do the work necessary to help aging populations.[33]

The Silicon Valley investor Marc Andreessen takes the most optimistic view about technological innovation. He says, "The job crisis we have in the U.S. is that we don't have enough workers," not that automation will reduce jobs. He believes that in much the same way that automobiles created new kinds of jobs, such as car repair, vehicle rentals, and parts dealerships, self-driving cars will create "a whole set of new jobs."[34]

THE VIEW OF THE PUBLIC

It is not just experts who debate the workforce ramifications. Public opinion polls show that the general U.S. population is paying attention to the possibility of automation-based job losses. A national survey of the general public revealed considerable unease about emerging trends. A Pew Research Center national public opinion survey found that 65 percent of American adults think that in fifty years, robots and computers "will do much of the work currently done by humans."[35] Most of these individuals are not pleased with this development. When asked for their assessment of this shift in the health care area, "65% think it would be a change for the worse if lifelike robots become the primary caregivers for the elderly and people in poor health."[36] People also are worried about the emerging technology of driverless cars. When asked in 2017 whether they would ride in a driverless car, 44 percent said they would while 56 percent said they would not.[37]

Most recently, a public opinion poll undertaken by Burson-Marsteller and PSB Survey found that more people believed advances in automation and machine intelligence would eliminate jobs than believed such advances would create jobs. When asked about this, 64 percent believed automation would eliminate a significant or moderate number of jobs, while only 18 percent thought it would create positions. Many people think, "Manufacturing jobs of the future will require knowledge of automated manufacturing systems and other advanced skills such as mechanical or electrical engineering."[38]

Another survey by the Pew Research Center found that "many Americans anticipate significant impacts from various automation technologies in the course of their lifetimes—from the widespread adoption of autonomous vehicles to the replacement of entire job categories with robot workers." In particular, "64% expect that people will have a hard time finding things to do with their lives if forced to compete with advanced robots and computers for jobs."[39] When asked if they were worried or enthusiastic about robot automation, 72 percent indicated they were worried while 33 percent said they were enthusiastic about the change.[40]

There were substantial differences by sector in people's concerns about job losses as a result of robots or computerization. As shown in figure 4-3, the greatest job impact is expected in the hospitality sector (42 percent), followed by retail and finance (41 percent each). In contrast, the lowest job impact is expected in education (18 percent) and health care (24 percent).[41]

Even business executives are sanguine about the job impact. When asked about their own hiring plans over the next five years in the context of increasing use of robotics, 58 percent of CEOs queried said they planned to reduce jobs, while only 16 percent said they would increase jobs.[42] Since CEOs are on the front lines of hiring, these outlooks do not bode well for workers. An A.T. Kearney survey of business leaders found that "the increasing sophistication of AI has helped dramati-

Figure 4-3 People's Perceptions about Sector Job Losses, 2017

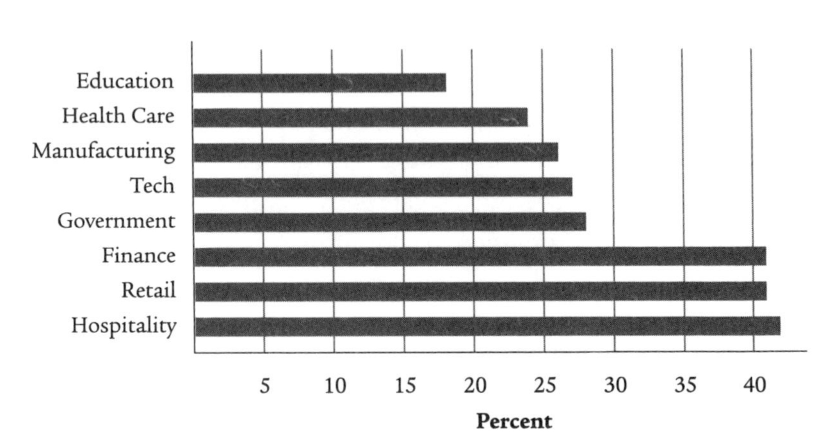

Source: Aaron Smith and Monica Anderson, "Automation in Everyday Life," Pew Research Center, October 4, 2017.

cally reduce the number of man-hours needed to sort through financial litigation and regulatory compliance."[43]

WORKFORCE DIFFERENCES BASED ON DEMOGRAPHY

The workforce ramifications discussed here are not likely to be equally distributed across all demographic groups. Rather, the impact differs based on age, gender, income, education, and race. Certain individuals are more at risk than others for unemployment as a result of the increasing use of emerging technologies. As the digital economy unfolds, it is clear that workers with few technical skills or other in-demand skills likely will face rough going in the future.[44]

Table 4-2 shows the national unemployment rate by demographic group in 2017. Overall the economy was doing well, with a 4.2 percent household unemployment rate. However, there were significant differences by age and race. About 14.7 percent of those aged sixteen to nineteen years old looking for work were unemployed. That figure rises to 25.9 percent for sixteen- to nineteen-year-old African Americans and 33.1 percent for sixteen- to nineteen-year-old male African Americans.[45]

Young people and racial minorities face particular risks from technological change as they are the ones whose job prospects are more likely to be affected by robotics, machine learning, and AI.[46] Even though many of them have time to acquire relevant expertise, few are getting training in science, technology, engineering, and math (STEM) fields. This limits their ability to withstand the coming disruptions. According to the U.S. Department of Education, there will be a 14 percentage-point increase in STEM jobs between 2010 and 2020. However, "only 16 percent of American high school seniors are proficient in mathematics and interested in a STEM career."[47]

In light of that sobering fact, the workforce impact of emerging technologies is likely to be profound. According to the economist Jef-

Table 4-2 National Unemployment Rate by Demographic Group, 2017

	Percent
Overall	4.2
White	3.7
African American	7.3
Men	4.2
Women	4.3
16–19 years of age	14.7
20–24 years of age	7.1
25–54 years of age	3.6
African Americans, 16–19 years	25.9
Male African Americans, 16–19 years	33.1
Male African Americans, 20–24 years	13.0
Female African Americans, 16–19 years	19.6
Female African Americans, 20–24 years	8.8

Source: Bureau of Labor Statistics, "Labor Force Statistics from Current Population Survey," June 2017.

frey Sachs, "Robots and artificial intelligence are likely to shift national income from all types of workers to capitalists and from the young to the old."[48] That will exacerbate income inequality, alter the lists of winners and losers, and widen social divisions. "An expanded economic pie favors those with managerial and professional skills who can navigate the complexities of finance, administration, management, and technological systems," he notes.[49]

In countries that have faced civil unrest or war, youth unemployment already is quite high and the societal consequences are quite dire. According to the World Bank, Libya has a youth unemployment rate of 48.1 percent, followed by 36.1 percent in Iraq, 33.4 percent in Egypt, 31.5 percent in Syria, and 26.6 percent in Algeria.[50] When there are few jobs for young people (or for people in general), some of them turn to crime, terrorism, or other forms of social or political discontent. Having a dismal economic future does not encourage social integration or societal peace. If future unemployment rises among prime-working-age men due to automation, there could be a substantial increase in criminality or social unrest in those places.

But job complications are not limited to young people. Women (and some men) have entered positions that focus on caregiving. With the aging population and the shift of jobs toward health care, that would appear to insulate people employed in those areas from technological change. Yet digital technology is changing caregiving. Sensors and remote monitoring devices record vital signs and electronically transmit them to health care providers. Wearable technologies keep people in touch with friends and family members. The sick and infirm no longer need just a human being to care for them but can access care through "intelligent family care assistants" that track their health and notify professional caregivers when problems intensify.[51]

One study calculated that 11 million seniors live by themselves in America. A number of these individuals use "an emergency alert system for the elderly based on monitoring of their heart rates, breathing activities, and room temperature measurements. The device also allows the dependents to make on-demand requests for assistance."[52] While these tools represent a small aspect of current caregiving, they are likely to grow in the future as new devices with enhanced capabilities become more prevalent.

Racial minorities face dismal job opportunities even in the best of times. Owing to discrimination, prejudice, and lack of IT-related training, minorities already have high unemployment rates. Moreover, without skills training, it will be difficult for them to adapt to the new economy when advanced machines take their jobs. They are likely to suffer disproportionately from emerging technologies.

Their ability to benefit from digital technology is limited by uneven access to computers and high-speed connectivity. An analysis of digital inequality shows that many of these individuals lack access to high-speed internet, and this creates difficulties in terms of education and employment.[53] They are less likely to own smart phones, have access to the internet at home, or get detailed instruction in computing software. That restricts their ability to adapt to the emerging workforce of the twenty-first century.

NEW BUSINESS MODELS AND THE SHARING ECONOMY

As digital disruptions unfold in many sectors, they are accompanied by major changes in business models. Over the past few decades, American firms have moved a number of their traditional functions outside the firm. In an effort to cut costs and achieve efficiencies, they rely on suppliers from around the world, third-party agents to handle benefit provision and claims processing, and temporary or external workers to handle worksite cleaning, accounting, and communications.

These changes have produced what Brandeis University's Professor David Weil calls the "fissured workplace."[54] Rather than manage operations on their own, business leaders have devolved authority to a broad network of outside companies, distant suppliers, and remote managers. The result has been a workplace that relies on external firms to handle company operations. Many of these entities cut costs, reduce the number of in-house employees with benefits, rely extensively on temporary workers, and weaken external oversight of workplace standards.

In the emerging economy, many workers have become independent contractors with few benefits. Firms recruit low-paid workers and rely on temporary employees or part-time staff. There is little loyalty between employers and employees, and companies "race to the bottom" in order to gain a competitive advantage. Negative consequences, such as higher income inequality or lax safety standards, are seen as "externalities" best left to society as a whole to manage. In many cases, these problems fester because no one has an incentive to take them very seriously.[55]

One of the best examples of structural change due to technology is the so-called sharing economy, defined as "the peer-to-peer-based activity of obtaining, giving, or sharing the access to goods and services, coordinated through community-based online services."[56] In the past, people thought of jobs as permanent, full-time positions. Workers who were employed at least thirty hours a week received benefits such as health care insurance, retirement funding, and disability insurance. They worked in the physical office of their employer and provided concrete goods or services.[57]

However, many of the jobs enabled by technology or mobile apps are temporary or episodic in nature.[58] People drive for Uber when they have time and are paid based on the number of rides they deliver. If drivers want to work ten or twenty hours a week, they do so, but they are paid commensurate with their workload. Those who work many hours can be well compensated while those who devote fewer hours will make considerably less money than others..

As a sign of the transformation already taking place, the ride-sharing business in the United States increased by 69 percent from 2010 to 2014 and by another 63 percent from 2014 to 2015.[59] The popularity of Uber and Lyft in car sharing, Capital Bikes in bike sharing, and Airbnb in room rentals suggests how business models have altered.

Similar trends have unfolded in China. According to news reports, "The State Information Center's Sharing Economy Research Center calculates that some 600 million Chinese conducted sharing economy business worth $500 billion in 2016, up 103 percent over 2015, with predictions the sharing economy will account for 10 percent of China's GDP by 2020."[60] Popular services there include car sharing, bike sharing, and apartment sharing. There even are entrepreneurs who allow consumers to share umbrellas, basketballs, and phone chargers.[61] With the popularity of mobile payment systems, it is easy to set up sharing services and allow customers to use smart phones to make payments. According to business people, "The payment systems integrate seamlessly with a user's bank account and allow even tiny transactions with simple taps and camera snaps."[62]

In the EU, "More than half of all new jobs created . . . since 2010 have been through temporary contracts." Overall, among young people, 40 percent hold short-term jobs without benefits. It is difficult to pay for full-time employees and their accompanying full social benefits, so firms rely on temporary workers, who do not require costly benefits. In Spain, with its debt crisis and poor economy, this trend is even more pronounced. Only 10 percent of the jobs (1.7 million positions) created in 2016 were permanent. Most (18 million) were temporary in nature.[63]

Despite these dire ramifications for the workforce, sharing services remain popular with the general public. As an example, Pew Research Center surveys show that "72 percent had used a shared or on-demand online service." The services in the survey included second-hand goods sites such as eBay and Craigslist (used by 50 percent of Americans), Amazon Prime delivery services (40 percent), Uber ride sharing (15 percent), room-sharing sites such as Airbnb (11 percent), temporary worker hire (4 percent), and renting products (2 percent).[64]

According to a report from the Freelancers Union, about one-third of the U.S. labor force (around 53 million Americans) who provide these sharing services are freelancers, that is, part-time workers without benefits. That figure is expected to increase to 40 percent by 2020.[65] This number includes those who work part-time (roughly 16 percent of the workforce) plus those in conventional employment who undertake projects on the side in order to earn a living.

This model works well for those who value flexibility and want part-time work. They may have alternative income from family members or may string together a series of part-time positions to make ends meet. In particular, young people who still receive support from their parents are able to eke out a living.

But jobs of this sort are problematic if they are a person's only source of income and the person receives no health or retirement benefits elsewhere. In such cases, workers struggle economically because there is no certainty about their employment. It is impossible to support a family on a part-time job, and most such positions provide no benefits (or only nonportable benefits) because the workers are considered independent contractors for the firm, not full-time employees.[66]

For this reason, some countries are considering new regulations over this part of the economy. In China, for example, authorities are encouraging the sharing economy but seek to address issues such as "market entry, fair competition, appropriate regulatory mechanisms, supportive government services, the importance of trust mechanisms, complementary legislative support, the protection of consumer and personal information rights and interests, labor relations, taxation and liability issues."[67] That nation's State Council has put forward policy guidance to coordinate the services of car-sharing services with taxi operators to ensure fairness in operations and taxation. It also is reviewing the rules on driver background checks and insurance requirements to make sure public safety is protected. These are the types of adjustment that many nations should consider as sharing services become more prevalent.

Other places in the West have witnessed worker unionization activities geared to the sharing economy. Uber drivers, for example, have concerns about minimum fares, tipping policies, and overall fares. In cities such as New York, San Francisco, Seattle, and Philadelphia, independent contractors have organized to contest company policies and demand better compensation.[68] It is unclear how successful these efforts will be, but it is worth watching as the sharing economy becomes more prevalent.

NEW JOBS THROUGH VOLUNTEERING AND PARENTING

In a situation in which a number of jobs are temporary and the social benefits are uncertain or nonexistent, it makes sense to broaden the definition of work to include part-time labor, volunteer activities, mentoring, and parenting. These are the kinds of positions that contribute to society but currently provide little income and no health benefits. People participate in community activities because they value the work of public-minded organizations. They help other individuals, train the next generation, or provide assistance for the less fortunate in society. But they are not officially counted by society as workers or valued in terms of social betterment except for occasional community awards as public-spirited volunteers.

This perspective is consistent with viewpoints developed by the Shift Commission on Work, Workers, and Technology. It convened a number of experts in cities across America and debated four different future employment scenarios: (1) more work, mostly jobs (similar to the current situation), (2) less work, mostly jobs (a recession scenario), (3) more work, mostly tasks (a sharing economy situation), and (4) less work, mostly tasks (a shift to volunteering, parenting, and mentoring).[69]

A variety of survey evidence demonstrates that young people are particularly interested in the fourth category of this typology. In general, millennials have different attitudes toward work and leisure time, and many want to pursue volunteer activities that contribute to the common good. For example, a survey of American students found that they want "a job that focuses on helping others and improving society." Quality of life is important, not just maximizing financial well-being.[70]

Many people value volunteer activities outside their working life. They have varied interests and want extracurricular activities that fulfill them. This may involve tutoring in after-school programs, helping English as a Second Language pupils, stopping domestic violence, protecting the environment, engaging in faith-based activities, or encouraging entrepreneurship. According to a Deloitte study,

"63 percent of Millennials donate to charities and 43 percent actively volunteer or are a member of a community organization."[71]

In a digital economy characterized by less work and more leisure time, we should think about certifying volunteer work toward eligibility for social benefits. In the United Kingdom, for example, volunteers are reimbursed for expenses or earn credits for job training programs through participation in worthy causes. In addition, volunteering counts as "looking for work" so people can use those activities to qualify for social insurance credits.[72]

Parenting is another activity that contributes to society but is unpaid under current job definitions. People provide care for babies and children (and sometimes parents or grandparents) but do not get compensated for this work. Parenting activities are crucial for society because every bit of research demonstrates that effective parenting and caregiving are crucial to life outcomes. Those who receive early care and nurturing are emotionally healthy, graduate from high school and college, and end up earning a good living. Conversely, those who do not are maladapted, more likely to be incarcerated, and have difficulty earning a living. Expanding the definition of work to include worthy activities such as these would not only recognize the contributions of, but would also provide an opportunity for, new kinds of jobs. This would benefit the overall community and gives people meaningful activities in which to engage, something that would be especially valuable in the transition to a new economy.

DEVELOPING LEISURE TIME THROUGH ART AND CULTURE

One possible benefit of new workforce trends is that people will have more leisure time than in the past. This can happen in one of two ways. Some people will not be needed in the new digital economy, so they will find other ways to construct meaning in their lives outside the workplace. Alternatively, even those who work may find themselves with time for other kinds of pursuits. Rather than most waking hours being spent on work-related tasks, the society of the future may have time for nonwork activities, including art, culture, music, sports, and theater.

The possibility of an end to work as we currently know it creates opportunities for personal enrichment. According to the Harvard economist Lawrence Katz, "It's possible that information technology and robots [will] eliminate traditional jobs and make possible a new artisanal economy . . . an economy geared around self-expression, where people would do artistic things with their time."[73] From his standpoint, this transition would move the world from one of consumption to one of creativity.

People will be able to use their leisure time to pursue a range of interests. Depending on their proclivities, they could have more time for family and friends. A study of family time found that macroeconomic conditions affect how much time people spend together. When employment problems rise, "fathers spend more time engaging in enriching childcare activities" and "mothers are less likely to work standard hours."[74] As long as there are opportunities for people to pursue broader interests, a reduction in work hours does not have to impoverish people.

Already, Americans undertake a variety of artistic activities. For example, a recent National Endowment for the Arts (NEA) Annual Survey found that 66 percent attended a visual or performing arts event in the past year, 61 percent consumed art through electronic devices, 45 percent personally performed or created art, and 43 percent read literature.[75] In general, "Women participate in the arts at higher rates than men across all categories."[76] The organization found that people had many motivations for arts participation. Seventy-three percent said it was to socialize with other people, 64 percent liked to learn new things, 63 percent said they wanted to experience high-quality art, and 51 percent wanted to support the community.[77]

A 2016 U.S. Census Bureau survey on public participation in the arts found a wide range of cultural activities engaged in by the general public. When asked what kinds of arts activities they had participated in in the preceding twelve months, 59 percent said they had gone to a movie, 44 percent indicated they had read a novel, 31 percent said they had danced socially, 21 percent said they had seen a theater performance, 14 percent had gone to a museum, 13 percent had knitted, crocheted, or weaved, 12 percent had played a musical instrument, 9 percent had gone to a classical music concert, 9 percent had sung with others, 8 percent had gone to a dance performance, 7 percent had gone to a jazz concert, 6 percent had read poetry, and 2 percent had gone to the opera.[78]

Many people pursue musical activities. They may sing for a church choir, play in a band, or play for friends and family members at informal gatherings. They like having a creative outlet and helping others enjoy the benefits of their talent. If anything, the development of digital music has accelerated the popularity of this area. According to Nielsen, "Music consumption is at an all-time high. Overall volume is up 3% over 2016, fueled by a 76% increase in on-demand audio streams."[79] The easy accessibility of digital music and the ubiquitousness of storage devices have accelerated this trend by making it easy, convenient, and inexpensive to access music and carry it on portable devices.

The growth of online audio encourages the digital distribution of music. Edison Research has tracked audio content over a number of years and found that the percentage of Americans listening to online radio has risen from 27 percent in 2010 to 57 percent 2016.[80] There has been a rise in terrestrial listening, but substantial increases in satellite and web-based content.

The same phenomenon is emerging in regard to the theater. This is an area that has a long and storied history in human civilization. From the ancient Greeks to Shakespearean audiences and Broadway theatergoers, many people have appreciated the opportunity to pursue dramatic arts or watch the performances of others. This is an area that a number find personally enriching, and it provides an outlet for artistic expression.

Some people are using their leisure time to develop hobbies, such as knitting, crocheting, rebuilding car engines, gardening, or woodworking. The Craft Yarn Council, which surveys knitters and crocheters about their art work, found that 65 percent said they liked doing it because it provided a creative outlet, 51 percent enjoyed making things for others, and 44 percent cited the sense of accomplishment from these activities. When asked to describe the benefits they felt, 93 percent reported a sense of accomplishment, 85 percent said it reduced stress, 68 percent noted that it improved their mood, and 56 percent felt it gave them more confidence.[81]

One of the pathologies of modern life has been an increase in sedentary lifestyles among many ordinary people. Lack of exercise and the high amount of "screen time" has produced a growth in the proportion of obese persons. Thirty-six percent of American adults and 17 percent of American young people are obese, according to the Centers for Disease Control and Prevention. Forty percent of middle-aged people are obese, and overall, obesity has increased among adults and young people.[82]

Concern over rising obesity has led to calls for greater participation in physical activities. One of the virtues of people's having more leisure time is enhanced opportunities for exercise and meditation. Membership in fitness centers and health clubs has increased from 32.8 million to 55 million Americans between 2000 and 2015.[83] In addition, participation in physical activity is up over the past two decades. According to the National Health Interview Survey, 49 percent of Americans engaged in regular aerobic activity in 2015, compared to 43 percent in 1997.

This was especially the case among young people. Almost 60 percent of those between eighteen and twenty-four years old said they did this, compared to 25 percent of seniors aged sixty-five or older.[84] Among the fastest-growing activities are high-impact aerobics, swimming, yoga, adventure racing, mountain biking, and triathlons.[85] Yoga and Pilates studios have also grown in popularity. The number of Americans practicing yoga grew from 20.4 million in 2012 to over 36 million in 2016. One-third of Americans say they have tried the Eastern regimen in the past six months.[86] These are particularly popular pursuits among young people.

Conclusion

Emerging technologies and major changes in business models have altered the manner in which people earn a living. They are affecting the nature of employment and undermining the traditional methods by which people accrue social benefits, especially those from underserved backgrounds. In the future, people should expect the pace of technological innovation to accelerate and to have a major impact on the overall economy. It is not that new jobs won't be created, but it is likely that older positions will be eliminated faster than new ones are created.

When workers with few skills are unable to find jobs, it is imperative that we broaden the conception of work to include such pursuits as part-time labor, volunteering, parenting, and mentoring. These activities enrich the overall community and help people develop identities outside their professional roles. It is important to do this because with many jobs being outsourced to independent contractors, more and more people are finding themselves in short-term positions that do not provide traditional health or retirement benefits. This will necessitate major changes to the social contract. Changing the benefits model now will improve our ability to help displaced workers who are negatively affected by technological innovation.

FIVE

A NEW SOCIAL CONTRACT

THE EMERGING ECONOMY presents challenges with respect to ensuring an income, health care, and retirement benefits. With employers moving toward greater use of temporary staffing with few benefits, it is vital that we as a nation figure out ways to provide essential services. Failing to come up with creative models risks a substantial increase in societal discontent. As noted by LinkedIn cofounder Reid Hoffman, "Transitions can be very painful. Let's try to make it work out in a way that's more humane."[1]

Yet despite the need for forward-looking thinking, there has been little public discussion of the societal impact of emerging technologies. It is crucial to understand the ramifications of knowledge societies and how they are exacerbating social and economic inequalities. In its most pointed form, the fear is that digital technologies will take away jobs, limit incomes, and expand the permanent underclass of unemployed or underemployed people. As argued by Nicolas Colin and Bruno Palier, "Employment is becoming less routine, less steady, and generally less well remunerated. Social policy will therefore have to cover the needs of not just [those] outside the labor market but even many inside it."[2]

If technology enables businesses to provide goods and services with fewer employees, what will that mean for wages and benefits? Under current policies, a significant increase in the number of people without full-time jobs would exacerbate socioeconomic divisions by weakening the distribution of benefits such as pensions, health insurance, and disability insurance. Since most benefits are tied to full-time employment, if the economy requires fewer workers, innovative models of benefits delivery will be needed.

In this chapter, I examine several alternatives for redesigning the social contract to address these problems. Possible provisions include establishing citizen accounts with portable benefits, providing paid family and parental leave, revamping the earned income tax credit to help the working poor, expanding trade adjustment assistance for technology disruptions, providing a universal basic income, and deregulation of licensing requirements so that it is easier to pursue part-time positions. Some combination of these policy adjustments will be needed to help people transition to a digital economy.

CITIZEN ACCOUNTS WITH PORTABLE BENEFITS

Right now, U.S. health insurance is a mix of public and private coverage. In 2017, 155.9 million Americans obtained health insurance through their employers, 62.3 million did so through Medicaid, 43.3 million through Medicare, 21.8 million through nongroup sources, 6.4 million through other public agencies, and 28.9 million (about 9 percent) were uninsured, according to the Kaiser Family Foundation.[3]

If unemployment or underemployment rises as a result of emerging technologies, an important workforce reform will be providing benefits and income in situations of non-full-time employment. Some experts estimate that up to half of jobs will be done by independent contractors by 2020, so there could be a growing number of people in this category.[4] In an economy characterized by temporary bouts of work, workers will need some means of getting essential benefits.

One way to deal with this dilemma is through the creation of what the policy analysts Colin Bradford and Roger Burkhardt call portable and flexible "citizen accounts." According to their formulation, each person would control "the current patchwork of funds for: education, training, health insurance, personal savings, retirement, life insurance, unemployment compensation and social support funds." The idea is that through this mechanism, "individuals would be empowered to manage the inevitable disruptions in their lives caused by job loss, re-training to adjust to technological change, re-location spurred by regional economic shocks, local market collapses, or by diminished economic growth."[5]

From their standpoint, this is a viable option for addressing employment difficulties associated with technological innovation. At a time when business models are changing fast and many workers lack the skills to transition to other positions, there needs to be a way for them to get benefits and retraining even when they are not permanently employed. Considerable money is currently being spent on worker assistance, but the programs are not coordinated, and employees have little say over the use of the funds. As a result, they do not gain the advantages of all the money being spent on their behalf.

A related variation is what writer Eli Lehrer proposes as "worker-controlled benefit exchanges." Rather than tying social benefits to jobs, this formulation would offer several ingredients of a social safety net, such as "unemployment insurance-like coverage, paid leave, some aspects of workers' comp insurance (for gig-economy platforms not taking part in the workers' comp system), health insurance, and other benefits."[6] The exchange would help displaced workers gain access to needed benefits and aid them until they found a new job. It would provide flexible benefits for those not in full-time employment and tide workers over during periods of unemployment.

Still another possibility is government-run benefit exchanges. An example is the Affordable Care Act, which extended insurance to people who previously lacked coverage. Whatever its merits as a health care reform initiative, its model of separating benefits from jobs was far-sighted in terms of the future economy. The legislation set up insurance exchanges in each state that sold health coverage even if someone had no full-time job. For those without the financial resources to pay for insurance, the federal government provided subsidies on a sliding scale linked to family income. As demonstrated by the 20 million Americans who gained health insurance through it and the accompanying expansion of Medicaid, this was a meaningful way to help those who could not get insurance through traditional jobs.[7]

Finally, the economists Seth Harris and Alan Krueger propose a new category of "independent worker" that bridges the gap between full-time employees with regular benefits and independent contractors with no benefits. According to their proposal, "Businesses [could] provide benefits and protections that employees currently receive without fully assuming the legal costs and risks of becoming an employer. Such benefits and protections include the freedom to organize and collectively bargain, the ability to pool (for example, a suite of employer-provided benefits such as health insurance and retirement accounts; income and payroll tax withholding), civil rights protections, and an opt-in program for workers' compensation insurance."[8]

Unlike full-time employees working at least thirty hours a week for a specific employer, independent workers would not be eligible for overtime pay, unemployment insurance, or a minimum wage. They would have the option of signing up for health care insurance. However, companies that failed to offer this kind of benefit would "pay a contribution equal to five percent of independent workers' earnings (net of commissions) to support health insurance subsidies in the exchange."[9] That would enable individual workers to purchase health care insurance themselves, financed by the corporate contribution.

Running through each of these proposals is the idea that benefit portability is a key to survival in the emerging economy. In the new digital economy, people are moving across employers and sectors of the economy, and these movements likely will accelerate in the future. As noted by the analysts Daniel Araya and Sunil Johal, "Introducing portable benefits for independent workers so that pension and health care benefits can be taken from gig to gig while requiring contributions from technology platforms that employ these workers" is an important feature.[10] In today's world, workers need this kind of benefit flexibility to survive in a working environment that is turbulent and chaotic.

PAID FAMILY AND MEDICAL LEAVE

Revamping the social contract involves deepening the appreciation for tasks such as parenting and caregiving that are not fully valued by American society today. Unlike nearly every other Western nation, the United States does not provide paid leave for parents needing to take care of newborn babies or elderly relatives. Rather, mothers and fathers must use sick time or take unpaid leave to care for the very young, the infirm, or the very old. France and Germany offer fourteen to sixteen weeks of fully paid time off, while the United Kingdom provides 90 percent for six weeks and a flat rate for up to thirty-two weeks.[11]

A proposal put together by the AEI-Brookings Working Group on Paid Family Leave advocated the adoption of a new paid leave policy in the United States. Developed by Aparna Mathur and Isabel Sawhill, it proposed at a minimum that Americans should have eight weeks of pay at 70 percent of their regular salary, with a cap of $600 per week. The group's rationale was that "63 percent of [U.S.] children now live in households in which all parents work."[12] For that reason, and because of the importance of balancing work and family life, the researchers felt it was vital to design and implement a more generous leave policy.

However, there were disagreements over the scope of the policy and how it should be financed. While a majority supported longer and more generous support, some preferred to limit the leave to low-income families. Their thinking was that low-income families were the ones most in need and therefore should be the centerpiece of the policy. The task force did agree the policy should be financed through a payroll tax and savings achieved elsewhere in the budget. With clear health and economic benefits resulting from paid leave, the researchers convincingly argue it is time for America to end its status as the only developed nation not to provide paid leave for family care.[13]

EARNED INCOME TAX CREDIT

In their pathbreaking 2014 book, *The Second Machine Age: Work, Progress, and Prosperity in a Time of Brilliant Technologies*, Erik Brynjolfsson and Andrew McAfee propose an expansion of the earned income tax credit (EITC) as a way to provide income support for the working poor during the transition to a digital economy.[14] For example, the current policy offers a tax credit of up to $6,143 for families with three or more children. As people make more money, the size of the credit drops, and it is phased out completely at income levels ranging from $40,000 to $55,000, depending on marital status and number of children.[15]

The goal of this expansion is to encourage people to work but make sure they have basic support for very low incomes. According to the Urban-Brookings Tax Policy Center, around 26 million households receive around $60 billion in tax refunds or reduced taxes. Data suggest that this policy made it possible for 6.5 million people to emerge from poverty.[16]

Harvard Law professor Cass Sunstein also supports the EITC. He claims that if properly devised, it would "reduce poverty, boost employment, improve the health of infants and mothers, and increase the likelihood that people would graduate from college".[17] Raising this credit by around 8 percent would yield major returns, he says.

Brookings Institution researchers Elizabeth Kneebone and Natalie Holmes describe the EITC as "one of the nation's most effective antipoverty programs." It has generated discernible gains in income, health, and education. According to them, "It has positive lasting effects for parents, who have shown longer-run earnings increases and better health outcomes. At the same time, their children exhibit a host of benefits, from better school performance and higher rates of college enrollment to more hours worked and higher incomes in adulthood."[18]

The tax credit has benefited metropolitan communities across the country. Data analysis demonstrates that "the credit creates local economic impacts equivalent to at least twice the amount of EITC dollars received."[19] Not only does that strengthen poor families, it helps the communities in which they live.

However, for the EITC to be effective during times of unemployment linked to technology disruption, it needs to be revamped. Right now, some needy people are not eligible because of low income limits. In addition, income transfers take place only once a year, at the time of tax refunds. If large numbers of people have little income, the EITC should be configured and made relevant to the needs of the broader groups of people being affected by economic dislocation.

Raising the income limit would help to address large-scale employment problems. With the high costs of housing, education, and health care and the looming threat of workforce disruptions, more people will be at risk. Even those with jobs may not earn enough to cover their expenses. In this situation, the EITC provides a means with a demonstrated track record to help people through difficult times.

In addition, making refunds or credits available on a quarterly basis would provide greater flexibility to poor families in need of assistance. Brookings scholar Alan Berube cites focus group research demonstrating that annual payments impose significant hardships on recipients.[20] Rather than helping them alleviate poverty, this type of payment schedule does not address the budgetary challenges that working people face. Having payouts made at more regular intervals would improve financially strapped households' situation.

The results of a Chicago pilot project suggest there are positive outcomes from more regular EITC payments. The city gave 343 public housing residents the chance to receive half their tax credit in quarterly payments. Subsequent evaluation revealed "increased economic security (fewer missed bills and late fees, less food insecurity), decreased borrowing (payday loans and loans from family and friends), higher capacity to afford child care and education or training, and decreased financial stress (the ability to make ends meet from week to week)."[21]

EXPANDING THE TRADE ADJUSTMENT ASSISTANCE (TAA) PROGRAM

For a number of decades, the U.S. federal government has sought to help workers in certain fields who were harmed by international trade agreements. If someone loses a job as a result of a trade deal, that individual qualifies for job retraining, income support, and job counseling. The idea is that it is the responsibility of the government to help these individuals if global economic issues have adversely affected them. Factory workers, service workers, seafood workers, and farmers can file a petition with the U.S. Department of Labor, which "investigates the facts behind the petition; it applies statutory criteria to determine whether foreign trade was an important cause of the threatened or actual job loss or wage reduction. If the Department grants the petition to certify the affected worker group, individual employees in the group may apply to their State Workforce Agency for TAA benefits and services."[22]

According to researchers, the TAA program has identified around 4.8 million American workers since 1974 who were harmed by global trade, or about "3 percent of mass layoffs [due] to import competition and relocation overseas."[23] Certified employees who are over the age of fifty years get wage supplements if they end up in a lower-wage position. This is relevant because a study found that "TAA participants earned 30 percent less on average than they made in their previous positions."[24]

However, the program could be improved. Right now, it is focused on one-time dislocations resulting from international trade deals. Beneficiaries have to prove that they lost their job because of the agreement as opposed to some other economic force. If they can demonstrate a detrimental impact, they qualify for government benefits.

There are no provisions, though, for economic dislocations that are currently emerging beyond trade agreements. For example, if workers lose positions as a result of automation or the use of robots, AI software, or autonomous vehicles, they are not eligible for government assistance through this program. They may be eligible for assistance under other programs, such as unemployment insurance, but not this one.

In addition, a relatively low number of individuals have been certified over the past four decades. As noted above, only 4.8 million workers have been certified under this program and just 2.2 million of them have actually received benefits.[25] This represents a thin slice of those affected by global commerce and business outsourcing. For this program to be helpful, it needs to cover other types of economic situations that produce mass layoffs or major dislocations. The model it has pioneered of helping workers through job retraining, income support, and job counseling is quite relevant to the current U.S. economic situation, but could be applied to more job dislocation circumstances than is currently the case.

PORTABLE RETIREMENT SUPPORT

Many employers have moved to 401(k)-style pension plans as a way to provide portable retirement benefits. Employees contribute to their own account and get a match from the employer. They control their own investment decisions and take the money with them if they switch jobs. Many organizations have generated positive results for their employees by automatically enrolling workers in retirement accounts.[26] When people are "nudged" to enroll in programs that benefit them, many more are likely to do so. Enrollees end up with more money and a more generously funded retirement plan.

The 401(k) saving program was originally designed in 1978 as a supplement to existing pension programs. But in many cases it has now replaced the traditional pension. In 1979, 38 percent of American private sector workers had a traditional pension, but that number has dropped to 13 percent of workers in 2017. Rather than rely on a benefit-defined alternative, the 401(k) plan has become a common vehicle for retirement saving.

However, only 55 percent of Americans currently have a retirement account, and the average person has only around $50,000 in overall retirement savings. This is far below what will be needed for actual retirement. According to Boston College's Center for Retirement Research, 52 percent of households are likely to run low on money during old age.[27]

In addition, this retirement option does not help those outside the workforce. Nearly all pensions and social security payments themselves are tied to employment. People who have not worked are not eligible for retirement benefits, so we need to figure out ways to take care of those people since their numbers are likely to expand in the emerging economy. Without some means of providing retirement income, it will be difficult to help people in their twilight years whose work has been disrupted by technological innovation.

There are a couple of proposals to boost retirement savings. Some suggest that enrollment in retirement programs be made mandatory as a way to increase participation and savings dollars. Advocates see that as an inexpensive way to boost retirement support by encouraging people to undertake actions that are in their own self-interest.

Others argue that companies should improve their matching gift programs as an inducement for people to save more. Not all firms offer a match, and failure to do so limits investment income and people's ability to finance their own retirement. Since businesses already provide retirement support, the argument is that expanding their matching programs is a way to help workers deal with income needs once they quit working.

The good news is that a number of firms already are increasing their retirement match. The average company has boosted its contribution from 3 percent in 2009 to 4.7 percent in 2017.[28] Reasons given include helping people save for retirement, improving worker recruitment and retention, and improving staff morale. Encouraging workers to prepare for retirement is a big boost for employees' long-term well-being.

Taken together, these initiatives represent ways to improve workers' prospects. But as business models shift to temporary hiring without benefits, questions remain as to how to enable senior citizens to live out their years with a reasonable quality of life. Finding a way to aid those individuals should be a high priority in the digital economy.

A UNIVERSAL BASIC INCOME

Because of the possibility of persistent unemployment or underemployment, some have suggested a universal basic income as a way to provide financial support for people in need. The economist Philippe Van Parijs has proposed that society "pay each citizen a basic income that would guarantee access to basic necessary goods."[29] That would help those with few employment prospects obtain basic subsistence without having to face homelessness or abject poverty.

Ben Schiller has written that "a universal basic income is the bipartisan solution to poverty we've been waiting for." He claims that with jobs disappearing to robotics and wages stagnating, governments should provide "a single payment that would give someone the chance to live reasonably."[30] The British economist Robert Skidelsky agrees that it is time for a basic income. He argues that "as robots increasingly replace human labor, humans will need incomes to replace wages from work." He claims that raising the minimum wage will not be effective because it will lower the cost differentials of labor versus machines and therefore speed automation.[31]

In the United States, the writer Charles Murray has outlined a more radical policy proposal. According to the policy writer Max Ehrenfreund, Murray's formulation "would amount to $10,000 a year for every adult citizen over the age of 21, along with an additional $3,000 dedicated to health insurance. . . . [But he] also proposes eliminating Social Security, Medicare, food stamps, housing assistance, and all the other programs the country has in place to help the needy."[32]

Other variations of a basic income would increase the monthly stipend to $2,000 per month (based on a U.S. poverty level of $24,000 per year) and retain some of the assistance programs. This is the more progressive version of Murray's plan in which the benefits are higher and the link to existing programs is clearer. Instead of disrupting the status quo by ending social welfare programs, as Murray proposes, this approach builds on current entitlements and extends them in novel ways.

Proponents see a basic income as a way to provide people with greater flexibility in social welfare support. In many countries, people who cannot find jobs are given unemployment compensation, but the amount is reduced if they earn other money. That provision creates a perverse incentive to not take on outside jobs for fear of jeopardizing the monthly checks. It is better to do nothing and take the government's money than to accept paid projects that endanger unemployment revenue.[33]

In addition, supporters like the security offered by an income guarantee. The RSA researchers Anthony Painter and Chris Thoung argue that a "basic income smooths work transitions whilst providing security in an age of potentially rapid technological change."[34] Since these experts believe there will be more periods in which people are unemployed and in need of job retraining, they see a guaranteed income as crucial to maintaining social mobility.

Critics of a basic income generally emphasize two factors. First, they point out the value that work adds to human worth. Many people derive a significant part of their self-esteem from their jobs. Even though a large number report they are unhappy in their current position, jobs are vital to many people. Second, critics of a basic income guarantee worry about the lack of work incentives in such a program. Much as with the controversies over welfare, commentators fear that people will stop working and contribute little to community betterment.

As an illustration, Rob Atkinson of the Information Technology and Innovation Foundation claims a basic income would "encourage people not to work and divert spending from activities that would create more jobs for people without jobs."[35] He also disagrees with the broader premise underlying a basic income, that technology will destroy jobs. According to his viewpoint, "No organization automates unless it saves money, and those savings get passed on to consumers, who in turn use those savings to buy something else. That spending creates jobs in other parts of the economy."[36]

Yet evidence from experiments abroad shows that giving people minimum support does not create dependency or lead to personal laziness. According to Charles Kenny of the Center for Global Development, providing a social safety net "may help lift people up and out of poverty. Give poor people cash without conditions attached, and it turns out they use it to buy goods and services that improve their lives and increase their future earnings potential."[37] Thus a basic income guarantee represents a way to help people who face difficult economic circumstances that are not of their own making.

To deal with dependency concerns, a basic income could be tied to volunteer activities or work requirements. Derek Thompson cites the Works Progress Administration example from the 1930s of having "the government to pay people to do something, rather than nothing."[38] He suggests the creation of a "national online marketplace of work" in which people could engage in projects that help the community. These activities could include tutoring, providing elder care, providing child care, aiding in disaster response, or arts and culture work. In this way individuals could contribute to the broader societal good while earning a minimum income from the government.

Finland is experimenting with local versions of a basic income. In a few communities, the government provides a monthly payment of 800 euros (around $900) in place of current social benefits. An alternative formulation pays 550 euros but the recipient retains existing income and housing support.[39] This income support is maintained even if an unemployed individual gets a job.

The Netherlands also has trial programs under way in the city of Utrecht. They provide welfare recipients with a minimum income and no questions asked "to determine whether a welfare system with no rules results in a happier, more productive society."[40] Scotland has launched an experiment in Fife and Glasgow that supporters see as a way to simplify government programs and promote solidarity. Council member Matt Kerr argues, "It says everyone is valued and the government will support you. It changes the relationship between the individual and the state."[41]

A town in Manitoba, Canada, saw positive results from a MINCOME guaranteed annual income. Comparisons of residents in the same town who participated in the program and those who did not found that "families receiving MINCOME had fewer hospitalizations, accidents and injuries" and "the high school completion rate ticked up." Most of the participants ended up above the poverty line, and few of them quit their jobs. The economist Evelyn Forget analyzed the data and concluded that "cash from the government eased families' economic anxiety, allowing them to invest in their health and plan over a longer horizon."[42]

So far, however, the general public remains skeptical of a guaranteed income. A 2016 Swiss referendum on the subject was decisively rejected by voters. It set the amount at 2,500 Swiss francs (or around $2,600 per month) for each adult and about 600 Swiss francs for children. Seventy-seven percent of voters rejected the initiative, while only 23 percent supported it.

An exit poll found that "the majority of voters rejected the proposal because they did not see it as financially feasible. They also cited concerns that unconditional income would attract more foreigners to Switzerland and diminish the incentive to work."[43] Those who supported it felt the proposal would promote new work and lifestyle models, and would give value to household and volunteer work. The exit poll indicated that 44 percent favored pilot projects in local communities to test the program and 49 percent did not.[44]

People's concerns about the cost of the basic income seem warranted, based on data from other countries. A study of the basic income in Australia estimated it would cost $340 billion a year, which is almost twice the $192 billion the national government currently spends on welfare and social security.[45] That is a large amount of money to pay for this type of income program.

With these types of public concerns about cost and effectiveness, it is clear that advocates need to do a much better job explaining their policy proposal if they wish to build future popular support. People have a number of questions about this idea and want to see better evidence it achieves the particular policy outcomes that are claimed. They are not ready to implement this idea on a widespread scale until they see concrete benefits derived from pilot projects.

JOB LICENSING REQUIREMENTS

One of the current barriers in the transition to a digital economy is professional licensing requirements. In the 1950s, according to the Institute for Justice, "only 5% of all jobs required licenses to practice. Today, almost a third do."[46] Most of these requirements are at the state or local level, so action is needed to make significant changes. As a general matter, fewer jobs should be required to have licenses, and the time and money needed to get a license should be reduced as long as it does not endanger public health or consumer well-being.[47]

These types of reforms would give people more flexibility to switch occupations and learn new skills. If they did not have to take expensive and time-consuming courses in order to qualify for a new position, that would facilitate job transitions. In an era of considerable dislocation and disruption, simpler professional job licensing requirements would help people cope with impending layoffs and employment changes. If properly tailored, these alterations should not endanger public safety or threaten community well-being. Instead, they would provide more flexible options for employment and make it easier for those of limited education to earn a living. Having more flexible work requirements would be beneficial to people during the transition to a new-style economy.[48]

WHO SHOULD PAY?

As with any revision of the social contract, it is important to consider how to pay for benefit enhancements during a time of economic dislocation. One technology billionaire, Bill Gates, gained considerable attention recently when he proposed a "robots tax" as a way to help displaced workers get needed job retraining. His argument was that robots are taking jobs in the workforce and therefore need to be part of the solution for those adversely affected by technological change. In an interview, he observed: "If a human worker does $50,000 of work in a factory, that income is taxed. If a robot comes in to do the same thing, you'd think that we'd tax the robot at a similar level."[49]

But reaction to his proposal largely has been negative. Former treasury secretary Lawrence Summers has accused Gates of "rejecting technological progress" and "singling out robots as job destroyers." The distinguished economist pointed out that technology promotes greater efficiency and therefore should not be discouraged through additional taxes.[50] Similarly, the Bloomberg columnist Noah Smith ridiculed the idea. "Imposing added costs on technology will slow growth and won't help people displaced by automation," he noted. The commentator correctly pointed out that the problem is not so much the technology but rather the inequality created by the technology. He reminded readers that "wages in Britain fell for four decades at the start of the Industrial Revolution," and that a negative impact of that sort is typical during periods of major economic change.[51]

Even though Gates's robot tax may not be the correct mechanism for dealing with the current economic transition, the billionaire surely is correct in noting the world faces a major retraining problem and existing programs do not come close to dealing with the scope of the workforce challenge. Unless we get serious about impending economic transformations, we may end up in a dire situation of widespread inequality, social conflict, political unrest, and a repressive government to deal with the resulting chaos.

A substantial part of digital disruption concerns economic inequity. The negative political and economic ramifications that flow from technology change involve how to help workers gain job skills when they lose positions and who will pay for the transition costs. There clearly are going to be substantial side effects associated with the shift to a digital economy. Women and minorities are likely to suffer from economic dislocations. In addition, middle-aged men and older employees will have a difficult time gaining needed skills and successfully competing for job positions.

There are several ways to pay for new social programs. One way is to raise income taxes on those earning over $466,000 per year, which has been the popular liberal approach in recent years because it focuses the tax increase on the top 1 percent of earners. For example, a 10 percent surcharge on the wealthy would raise funds that could be targeted toward worker retraining or new benefit programs. It is a way to tax those who have done very well economically over the past several decades so that they partially finance those who have not done very well. As of 2017, the top 1 percent comprised around 1.4 million taxpayers, and they paid $542.6 billion in federal income taxes. At current rates, a 10 percent surcharge would generate around $54 billion in new revenue each year, assuming no major increase in tax avoidance strategies.[52]

This type of approach, though, is not likely to raise the amount of money needed to deal with job losses linked to technology change. The resources required are much greater than the question of whether the wealthy pay a 37 percent, 39.6 percent, or 44 percent marginal tax rate on their incomes. None of those tax rates would generate the resources needed to deal with widespread economic dislocations.

Another possibility is a progressive tax on high consumption goods. Gates has suggested this as an option to deal with inequality. He believes rich individuals should pay more than they currently do through income taxes. He focuses on a "high consumption" tax because that is targeted to the ultra-wealthy and taxes them extra only when they spend large amounts of money on large homes, yachts, expensive automobiles, and the like.[53]

Still another policy alternative is a solidarity tax to address economic dislocations. This is a tax on the net property, stock, pension, and financial assets owned by high-net-worth individuals.[54] A number of countries already do this. Since 1982, France has had a "solidarity tax" of 0.5 to 1.5 percent on net assets over 1.3 million euros (around $1.5 million).[55] Norway taxes net assets over 1.2 million kronas at the rate of 0.85 percent. Spain has a Patrimonio tax of 0.2 to 3.75 percent on net assets over 700,000 euros. Argentina taxes net assets of 800,000 ARS at a rate of 0.5 percent.[56]

Although it is likely he no longer would support this idea, billionaire Donald Trump proposed in 1999 a "one-time" wealth tax of 14.25 percent on people with a net worth of $10 million or more to help pay off the national debt. He claimed at the time it would raise $5.7 trillion and would go a long way toward putting the country back on the path to fiscal solvency.[57] His proposal went nowhere, but it did represent a bold initiative for dealing with the kind of structural economic change that is likely to arise in coming years.

In the United States, a 1 percent solidarity tax on net personal assets over $8 million would fall on households in the top 1 percent of the United States, according to the Urban Institute. Its economists have estimated that these households have at least $7.9 million in wealth, as of 2013.[58] If there were no exclusions for charitable contributions or other considerations, a 1 percent wealth tax would generate about $379 billion in government revenue each year.

This assumes the tax falls on the top 1 percent holding at least $8 million and that these individuals own 40 percent of the family wealth in America. The latter figure is based on Congressional Budget Office numbers estimating that the top 1 percent in 2013 held between 38 percent and 42 percent of family wealth. The exact percentage varies with particular methods used to calculate it. Data from the Survey of Consumer Finances, for example, put the percentage at 38 percent, while estimates by the economists Emmanuel Saez and Gabriel Zucman place it at 42 percent.[59] According to the Federal Reserve Bank, 2017 household net worth totaled $94.8 trillion.[60]

Monies generated by such a levy could fund hardship payments or retraining necessitated by economic restructuring and technological innovation. It could cover things such as citizen accounts, lifetime learning funds, an expansion of the EITC, and paid family and medical leave. It would generate funds from those Americans in the strongest position to pay, while leaving 99 percent of households untouched by the tax. Not only would it provide needed funding for important social programs, it would reduce the widespread economic inequality that has arisen in the United States.

Supporters of the proposal claim that a solidarity tax would help ordinary workers, reduce inequality, and help deal with the deleterious consequences of wealth for the political process. Daniel Altman, an economist at New York University, says, "Wealth inequality and lack of access to opportunity [are] destroying the meritocratic aspects of our economy. That will cost us growth in the long run." He wants to replace the income tax with a graduated tax on wealth. According to his formulation, "No tax might be imposed for a household's first $500,000 in wealth, 1 percent for the next $500,000 and 2 percent for wealth above $1 million."[61] Those revenues could be used to finance needed investments arising from economic restructuring and technology innovation.

One approach that will not work is cutting the tax rates paid by the wealthy as a way to stimulate overall economic growth. There is little evidence that this perspective, long a pillar of Republican policies, seriously addresses economic dislocation. Based on the policy experience of the last few decades, when inequality has increased and economic growth has decreased, massive personal tax cuts largely go to the top 1 percent and do not address the needs of the working class at a time of accelerating technological change.

An illustration of the pitfalls of this approach can be seen in the economic program of President Trump. He promised "to bring down taxes, dropping the rates on individual and corporate income, throwing out the estate tax and simplifying the system for ordinary taxpayers overall" and delivered on those goals with his 2017 tax cut. However, an analysis of his approach by the bipartisan Tax Policy Center found that its economic benefits "are overwhelmingly concentrated among the very richest taxpayers. Nearly half of the total savings (49 percent) would accrue to the richest 1 percent of households."[62]

The same flaw applies to GOP efforts to reduce corporate taxes. The Trump proposal for substantial tax rate cuts went substantially to the well-to-do. For example, an analysis by the Center on Budget and Policy Priorities found that "about 70 percent of the benefit of a corporate rate cut flows to the top fifth of households—with one-third flowing to the top 1 percent alone."[63] This type of fiscal approach is not effective at addressing the economic equity problems likely to be made worse by accelerating technology innovation.

CONCLUSION

The emerging economy presents challenges in terms of ensuring workers' income and social benefits. As employers move away from full-time jobs with benefits to temporary positions without benefits, it is vital that we figure out ways to support essential services. Emerging technologies allow businesses to provide goods and services with far fewer employees, so it is crucial to develop new models of benefit delivery.

This chapter has looked at several ways to redesign the social contract and pay for the benefits. Among them are creating citizen accounts with benefits outside of jobs, offering paid family and parental leave, revamping the EITC, providing a universal basic income, changing licensing requirements to ease employment transitions, and enacting a solidarity tax on the top 1 percent of wealthy Americans. Some combination of these proposals is needed to help people make the transition to a digital economy.

SIX
LIFETIME LEARNING

IN A WORLD of rapid technological, organizational, and economic transition, it is imperative that people engage in lifelong learning. The traditional model, in which people focus their learning on the years before age twenty-five, then get a job and devote little attention to education thereafter, is obsolete. In the contemporary world, people can expect to switch jobs, see whole sectors disrupted, and need to develop additional skills as a result of economic shifts. The type of work they do at age thirty likely will be substantially different from what they do at ages forty, fifty, or sixty.

For this reason, it is vital that people develop new capabilities throughout their lives. People need to stay abreast of the latest developments and understand that employers look for different things at various times in an employee's lifetime. Skills that might be perfectly suited for a certain period may become irrelevant and thereby force individuals to update their abilities for a changing workforce.

In this chapter, I examine a number of different vehicles for lifetime education. Community colleges, private businesses, and distance learning have vital roles to play in workforce development because of the need for vocational training that is inexpensive and accessible to adults. However, it is important that vocational approaches provide skills that serve an individual over the course of many years, not just fulfill an immediate need. I also discuss the need for curricular reform, whereby schools would provide the most relevant training for young people. Those preparing to enter the workforce need schooling that puts them on a solid course for a long period of time. Finally, I discuss lifetime learning accounts as a means to fund job retraining and continuing education courses. In a period of rapid change, people need some way to pay for new skills acquisition.

AN ERA OF PERMANENT DISLOCATION

As described in earlier chapters, the coming decades are likely to be an era of permanent economic dislocation. Major changes associated with technological innovation and the emergence of new business models are in the offing; some already are in place. The sharing economy is accelerating, and in the future more jobs will be temporary or episodic in nature.

These dislocations are expected to have a dramatic impact on the workforce and to increase the need for continuing education. One study has estimated that "65 percent of children in grade school today are predicted to work in jobs that have yet to be invented."[1] Those who are starting out in life or are early in their professional careers are likely to face employment instability and volatility. Disruption will be the hallmark of the future workforce.

There are several reasons for this eventuality. In particular, the use of digital technology is rising and is expected to penetrate every sector. We have already seen some of the effects in communications, finance, and entertainment. People get information in radically new ways. Businesses act on this material in new and creative ways. There may be a transition from text to video as the predominant form of mass communication. Social media allow anyone who is so inclined to become a digital publisher or videographer.

Massive changes are likely to occur in additional sectors in the coming years. Health care and education represent parts of the economy that have been least disrupted by digital technology. It has been challenging to innovate in fields such as these where skilled labor is most important and difficult to automate. Both areas are labor-intensive and highly regulated, and therefore are not very susceptible to productivity-enhancing innovations.

Yet even in health care, technology is being increasingly used in medical diagnosis and treatment. Data analytics and physician-assisted software are changing how caregivers practice medicine and deliver health care. The growing use of social robots will remake medical service delivery and patient care. There will be a variety of ways in which health care is altered through digital technology.

The same is true for education. Distance learning, massive open online courses, and digital resources are changing the way in which teachers educate and students learn. Young people have grown up as "digital natives" and are at ease using technology in many aspects of their lives. They now expect schools and colleges to offer the latest digital tools and to incorporate their use into the curriculum.

Some educational programs are innovating through games. Video designers are incorporating math or science puzzles into instructional games. Players are required to answer substantive questions in order to advance through the video game. Those who are successful earn points and win the game. This combination of video gaming and education represents a potent form of digital learning.

In addition, smart phones and tablets are changing how people get information and engage in a variety of activities. Mobile devices bring information and transactions to people's fingertips twenty-four hours a day. Through their convenience and accessibility, these products make it easy to engage in activities that used to require visits to physical facilities.

While these and other emerging technologies offer a number of important societal benefits, they are likely to be disruptive in terms of the overall workforce. Some of these initiatives will destroy more jobs than they create. Fewer truck drivers, restaurant workers, and retail clerks will be needed in the future, for example, even though these jobs have always been a mainstay for persons entering the workforce with a high school education. With some specialized coursework, they could get a job and earn a living doing other things. But they will need continuing education so that they can update their skills for other positions.

THE ROLE OF COMMUNITY COLLEGES AND PRIVATE BUSINESSES IN WORKFORCE DEVELOPMENT

Community colleges are vital in the contemporary situation because they train many adults who need coursework. With their lower cost and practical orientation, they are a venue of choice for people of limited means, immigrants, and working-class adults wanting to develop new skills. Since they are important in a workforce undergoing transition, it is vital that they be adequately funded so that they can fulfill their mission.

Vocational education and apprenticeship programs bring students closer to the current needs of the labor market. They help smooth young people's transition into the workforce or new careers. Students in these programs are able to enter the workforce with the particular skills that are needed and so can contribute to the business right away. In Colorado, for example, 181,000 students have enrolled in career and technical education (CTE) courses. They receive training in such subjects as criminal justice, agriculture, information technology (IT), and fashion design. According to program administrators, "94 percent of all CTE finishers obtained a job."[2] Analysis of community college programs has found that those with a clear tie to industry and that work to make sure the workforce development connects to company needs are the most effective ones.[3]

But it is not just educational institutions that are redefining their missions. Private companies also are embracing lifelong learning and worker retraining because of the difficulty of filling certain positions. A Deloitte survey found that "39% of large company executives said they were either 'barely able' or 'unable' to find the talent their firms required."[4] For technical companies, it is especially difficult to hire people with the requisite skills because many completed their education before the high-tech era or did not get training in digital areas.

To deal with this problem, companies have developed their own training programs and work closely with educational institutions to help workers learn new skills. John Donovan, the chief strategy officer and a group president at AT&T, says that half of the firm's workforce is "actively engaging in acquiring skills for newly created roles."[5] On average, his employees change roles every four years and therefore need new training. According to him, this is the new reality, and both companies and workers should embrace continuing education.

The economist Harry Holzer has analyzed the impact of automation and argues that the greatest risk facing workers today is "skill-biased technical change"—that is, workplace changes entailing the replacement of humans performing low-skilled tasks with automation and a commensurate increase in demand for better-educated workers to perform more complex tasks or those requiring social interactions.[6] In this situation, it is vital that workers receive the retraining that will give them the skills they need to remain in the workplace.

The good news is that many Americans today have embraced retraining. According to a survey undertaken by the Pew Research Center, "73% of adults consider themselves lifelong learners." These individuals have taken adult education courses or otherwise sought to upgrade their skills. According to the study director, John Horrigan, "Most Americans are interested in gaining new knowledge and skills—sometimes driven by anxiety about their jobs and often times spurred by the satisfaction of mastering new things and finding new ways to be helpful in their communities."[7]

There were several reasons for the interest in continued learning. Fifty-five percent of employed adults said they wanted "to maintain or improve their job skills," 36 percent sought "to get a license or certification they needed for their job," and 24 percent indicated they wanted to "get a raise or promotion at work." Thirteen percent said they did their training to find another job.[8] People already are seeing the need to upgrade their skills and adjust to new economic realities.

However, it is important that new skills help the individual over a period of years, not just on a short-term basis. As pointed out by Eric Hanushek and Ludger Woessmann, apprenticeship programs "facilitate the transition into the labor market but later on become obsolete at a faster rate."[9] The skills gained in apprenticeship programs and in some retraining programs are fungible. They help the individual over the short run but do not position the person for long-term economic success. It is important not to confuse temporary skills with the bedrock capabilities needed throughout an individual's lifetime.

DISTANCE LEARNING

Digital technology offers a host of possibilities for connecting far-flung or nontraditional students with the classroom. It brings geographically disparate students together with nonlocal instructors, in this way creating a richer variety of educational resources. It enables those who are remote from traditional institutions to take classes and gain access to various types of instructional materials.[10]

There are several different types of distance learning. Researchers distinguish between those that are web facilitated (up to 30 percent of course content is based on the internet), blended or hybrid (with online course content ranging from 30 to 79 percent), and fully online (with online content running at 80 percent or above).[11]

Most institutions offer blended offerings. And though many institutions advertise online courses, a number of them feature instructional techniques grounded in old-style pedagogical techniques. For example, the courses may have mechanisms for interactive chats but the discussion format itself may remain static and text-based. It has been difficult for institutions of higher learning to fully embrace technology and transform the manner in which they deliver educational content. Many schools graft technology onto their existing business models and instructional approaches, and so do not explore distance learning techniques for their transformative possibilities.

There is little doubt that distance learning has become a high-growth industry. A U.S. Department of Education survey of distance learning found that two-thirds of postsecondary colleges and universities in America offered "online, hybrid/blended online, or other distance education courses."[12] More and more people sign up for these courses and view them as important for their career development.

Higher education surveys conducted by Elaine Allen and Jeff Seaman of Babson University, for example, found that over 6 million college students in 2017 had taken an online course, up from 1.6 million in 2002.[13] This means that 29.7 percent of postsecondary students had engaged in distance learning, compared to just 9.6 percent who did so in 2002. Most of this digital instruction takes place at institutions of higher learning with enrollments in excess of 15,000 students. Very little online instruction is occurring among smaller schools with enrollments below that level.

A meta-analysis reviewed fifty studies that used experimental or quasi-experimental designs to examine the impact of distance learning on student learning. In general, the meta-analysis found that "students in online learning conditions performed modestly better than those receiving face-to-face instruction." Often, the study noted, online education had additional learning time that facilitated student instruction. These effects persisted for "different content and learner types."[14]

With student access to smart phones having tripled in recent years at the elementary and secondary level, the mobile platform for distance learning has risen dramatically. A Project Tomorrow survey of 350,000 K–12 students, parents, and administrators found that "62 percent of parents would purchase a mobile device for their child if their school incorporated them for educational purposes, and some 74 percent of administrators now believe that mobile devices can increase student engagement in school and learning."[15]

However, research studies have found a more complex outcome suggesting that what students learn varies across different metrics. For example, a study of an online teacher education program at the University of Waikato in New Zealand found it had a positive impact on dialogue creation but a negative one on learner autonomy. The "virtual classroom enable[d] users to interact using audio, video, and text and to share files, resources, and presentations using applications such as PowerPoint and Flash." Students could text one another, share a whiteboard, and see one another through webcams.[16]

After interviewing participants, researchers found that the online classroom "helped build trust and rapport and went some way toward developing a sense of identification with others in the group." Being able to see and hear one another in real time and interact online helped students come closer to the visual and audio experience of face-to-face instruction, and students liked that part of the virtual experience. About half the students felt the virtual classroom had contributed to their knowledge development, but many felt the technology did not encourage autonomous learning because class presentations were highly structured.

Daphne Koller of Stanford University teaches statistics through a combination of online and face-to-face interaction. Through the web, she presents video material with online questions that appear every five to seven minutes. Once a week, mandatory quizzes seek to keep students on track with the statistical material. Students can interact with each other and the teaching staff through an online discussion forum. They can pose questions whose answers can be viewed by anyone in the course. In-class activities focus on high-level discussions and real-world applications of mathematical information.

Her course surveys found that students liked "shorter chunks [of material], with rapidly moving content." When asked about the in-video quizzes, 72 percent of the fifty-six students described them as very useful, 24 percent thought they were fairly useful, and 4 percent found them irritating. Most also thought they came up at the right rate of speed and that the interactive sessions were useful. Based on these reactions, Koller concluded that online statistical education "can induce students to interact with the material during learning, with immediate feedback" and that "retrieval and testing significantly enhances learning."[17]

In focusing on online videos, this statistics course mirrors the educational success of the Khan Academy. The academy makes available 2,300 video presentations of various topics in STEM fields. It uses short videos (generally twelve minutes long) that are designed to accommodate students' short attention spans and present information in chunks that combine to form larger learning modules. Students can go through the modules at their own pace, and multiple-choice quizzes test them at different points during the learning process. Academy founder Salman Khan claims to have provided more than 54 million individual lessons through his videos. He says, "We're seeing 70 percent on average improvement on the pre-algebra topics in those classrooms. It definitely tells us it's not derailing anything. All the indicators say that something profound looks like it's happening."[18]

CURRICULAR REFORM

In today's world, it is important that schools train students for new jobs that will develop in the future. As the economist Andrew McAfee argues, "Our education system is in need of an overhaul. It is frustrating that our primary education system is doing a pretty good job at turning out the kinds of workers we needed 50 years ago. Basic skills, the ability to follow instructions, execute defined tasks with some level of consistency and reliability."[19]

What is needed, he said, are people who can do "things like negotiate, provide loving and compassionate care, motivate a team of people, design a great experience, realize what people want or need, [and] figure out the next problem to work on and how to solve it."[20] That is a radically different vision of education from what exists today.

A study of the future of work in the United Kingdom found there is a "shrinking middle" in the workforce that requires retraining. These are individuals in midlife or older who get laid off and generally do not have the skills required for other positions. Left to their own devices, they will struggle economically and suffer long-term unemployment. According to the research, "People moving in and out of learning will continue. In particular, when people develop portfolio careers, they need to be able to convert their qualifications or build upon the ones they have. Education has to come up with the right package to solve these new demands."[21]

With the fast pace of technological change and the development of new positions in data analytics or software coding, educational institutions that focus on traditional curricula are not providing young people with the skills needed in the twenty-first-century economy. They are training young people for the jobs of the past, not the future. As the educator Thomas Arnett has written, "Technology can multiply a teacher's capacity for differentiated instruction. Many adaptive learning platforms not only tailor computer-based learning experiences to students' individual learning needs, but also provide teachers with real-time, actionable data to help them intervene with struggling students."[22]

Curricula need to be restructured to focus on twenty-first-century skills. For example, collaboration and teamwork should be emphasized. Many contemporary positions involve working as part of teams, so it is vital that people learn those skills. In addition, it is crucial that people understand how to think critically and communicate their ideas to other people. If educational programs provide these types of skills, it will help students in an era of extensive digital innovation.

Both schools and universities need a closer alignment of curricula and the skills required in the workforce. They should develop courses that emphasize practical job skills. As argued by McAfee, young people have to develop a capacity for negotiation, communication, data analysis, and working effectively with others. Those talents are in short supply but high demand in the new economy.

Speaking at an education symposium, Mark Schneiderman, the senior director of education policy for the Software and Information Industry Association, said, "The factory model that we've used to meet the needs of the average student in a mass production way for years is no longer meeting the needs of each student." Instead he called for changes to education that would recognize the magnitude of the information changes that have taken place in American society, especially with young people. In today's world, he noted, students "are surrounded by a personalized and engaging world outside of the school, but they're unplugging not only their technology, but their minds and their passions too often, when they enter into our schools."[23]

As he pointed out, sticking to a twentieth-century production model makes little sense when twenty-first-century technologies are available. The key for educators is to figure out how to use technology to engage students in the same way that iTunes engages the aural sense and YouTube attracts the visual sense. The technology is at hand for education to become personalized and adapted to individual needs, but its adoption needs to be extended throughout the learning process.[24]

Writing many years ago, the psychologist Howard Gardner identified seven different types of intelligence: linguistic, logical-mathematical, musical, kinesthetic, spatial, interpersonal, and intrapersonal.[25] Formal education that focuses merely on intellectual ability as verified through IQ testing will miss the artistic, cultural, spatial, and emotional intelligences that exist in many people. According to Gardner, "Seven kinds of intelligence would allow seven ways to teach, rather than one."[26]

Wired classrooms and electronic instructional sets build on Gardner's insight by letting pupils learn at their own pace and in their own manner. Personalization makes education more adaptive and timely from the student's standpoint and increases the odds of pupil engagement and mastery of important concepts. It frees teachers from routine tasks and gives them more time to serve as instructional coaches for students.[27]

Rather than hewing to rigid time schedules and annual grade promotion, new curricula put students in control of their learning. Timewise, these curricular innovations are flexible and give students access to instructional material around the clock. In conjunction with teacher guidance, students undertake lessons at their own pace and based on their own preferred learning approach. As they master key concepts, they advance to higher levels of skills.

Often, instruction is based on specific projects that are of relevance to the individual, so that the topic is engaging to the student. One of the best indicators of learning is student engagement, so the more engaged pupils are, the more apt they are to learn the material. Those who are not interested in the content or the learning approach are not likely to be successful academically.

Many schools are being wired for high-speed data communications so that pupils can take advantage of technology that tailors the learning process. The New York City School of One represents a novel case of digital innovation in the classroom. Rather than having a single teacher for a specified group of students, this school employs team teaching targeted toward educating individual students. Each pupil gets a daily "playlist" with a variety of instructional activities geared to her needs. The playlist may include spending time with a teacher, watching an online tutorial, playing a video game, or using various types of electronic resources. Progress is tracked electronically, and students move to the next level when they have demonstrated appropriate skill mastery.[28]

The virtue of this approach is that it puts students at the center of the education process. Their daily activities are based on how they like to learn and what approaches deliver the best results for them. Pupils can receive instruction either one-on-one or in small groups of students. With computers tracking how they progress, instruction can be sped up or slowed, depending on the needs of the individual student. As a School of One student put it, "If I don't understand something, I can try and learn it in a new way and take my time. I don't have to learn it the same way everyone else does."[29]

Another promising program is High Tech High, which focuses on "personalization, adult world connection, and common intellectual mission."[30] It works with inner-city high schools that employ school-to-work strategies based on internships, fieldwork, and project-based assignments. Students are given a "staff adviser" who coordinates the individual's personal and professional development and works with family members. School members have access to laptops, networked classrooms with fast broadband, project rooms, and exhibition spaces.

Students have a mandatory work commitment that requires them to spend a semester interning with a local business or government department. School officials encourage students to have lunches with adults with a record of accomplishment and to participate in "shadowing" activities with outside mentors. This integration of work and school helps keep students on track and focused on what they want to do after graduation.[31]

The for-profit K–12 company enrolls around 81,000 students in twenty-seven different states in online education. Students "study on their own, clicking on lessons, doing exercises, taking tests, with teachers available by e-mail and phone for support." In this kind of independent environment, pupils must be self-motivated and able to work on their own.[32]

At the college level, instructors in some schools use a "backchannel" system called "Hot Seat." It provides a digital platform for students to raise questions or make comments during class discussions. One instructor found that it was a terrific way to get quiet kids more involved in the classroom dialogue. "It's clear to me that absent this kind of social media interaction, there are things students think about that normally they'd never say," explained personal finance professor Sugato Chakravarty. Before the software system, he noted, "I could never get people to speak up. Everybody's intimidated."[33]

The problem with past efforts at education reform is that many of them focused on raising performance but did not alter the manner in which instruction was offered.[34] The basic structure of the classroom stayed the same, with teachers presenting information and students taking tests periodically to demonstrate mastery. When little effort is made to alter the fundamental model and approach by which education takes place, it is difficult for students, teachers, and administrators to perform differently or raise levels of school achievement.

ACTIVITY ACCOUNTS FOR LIFETIME LEARNING

One way to encourage lifelong learning is through the establishment of what are called activity accounts. In an era of fast technological innovation and rapid job displacement, there needs to be a way for people to gain new skills throughout their working lifetime. When people are employed, their companies could contribute a set amount to an individual's fund. This account could be augmented by contributions from the person him- or herself. Similar to a retirement account, money in the activity fund could be invested tax-free in investment options, including cash reserves, stocks, or bonds. The owner of the account could draw on it to finance learning and job retraining expenses. The account would be portable, so that if the person moved or switched jobs, the account would migrate with that individual.

The goal of this account is to provide financing for continuing education. Under virtually any scenario, people are going to have to extend their education beyond the first two decades of their lives. Emerging jobs will need different skills from what people gained in school or college. New jobs will be created that do not exist today. As pointed out by the Brookings Institution scholar Kemal Derviş, as technological innovation continues, it will be crucial to provide people with a means to upgrade their skills and knowledge levels.[35] He notes that France already has established "individual activity accounts" that provide social benefits of this sort.

Scholars at the Aspen Institute have suggested a related idea they call "lifelong learning and training accounts." Modeled after individual retirement accounts, people can contribute up to $1,000 per year that is tax deductible with a match from the federal government or private employers. That would enable workers to finance retraining and adult education at any point up until their retirement. The goal is to create better-trained workers, a greater capacity to get jobs, and more flexibility in career transitions.[36]

Others have suggested a universal displaced worker program that helps those losing jobs develop new workforce skills. Proposed by the Obama administration, this fund would provide up to $4,000 a year for unemployed workers seeking to gain new training. Participants would not only receive limited financial support, they would get a stipend for child care, transportation, and some wage insurance if they gained a new job with lower pay than their previous position.[37]

Another idea comes from Peter McClure and is based on what he calls a "grubstake." Modeled on the GI Bill, this would be a grant from the federal government for all citizens when they turned eighteen years old.[38] They could use this money to pay for courses, go to college, or enroll in vocational programs. The purpose would be to promote higher education by providing money directly to the student. It would free students to select the program that benefited them the most.

Some states already have enacted free tuition programs for in-state students at their public universities. Michigan adopted this type of program for families earning up to $65,000. It is designed to improve access to higher education and generate a better-trained workforce.[39] New York has a similar program for families making up to $100,000. Oregon and Tennessee also provide free schooling at community colleges.[40] The thought behind this initiative is that society benefits from people who are more educated and that investing in the human capital of young people produces a general benefit for society and country.

Whatever the particular approach, adults will need financial support for continued learning. We should not envision education merely as a way for young people to learn new skills or pursue areas of interest. Instead, we need to think about education as an ongoing activity that broadens people's horizons and expands their interests and skills over the course of their entire lifetime. Education is an enrichment activity, and we need to view it as a continual benefit for the individual as well as for society as a whole.

CONCLUSION

In a world of rapid change, it is imperative that people engage in lifelong learning. In the contemporary world, people switch jobs, eke out a living in the gig economy, see whole sectors disrupted, and need new skills as a result. The type of work they do at age thirty is likely to be substantially different from the type of work they do later in life. As a result, it is crucial that they keep learning and develop new capabilities across their working lifetime. Failure to do so likely risks unemployment and undermines their economic future.

Community colleges and private businesses can help on this front, as can distance learning programs and options for personalized learning. Each of these options makes continuing education affordable and accessible to those who need retraining. They help provide new skills, while doing so in a practical manner. Some of the options are affordable, which helps those in need of retraining access the courses.

Lifetime learning accounts represent a way to fund job retraining and continuing education activities.[41] They enable people to upgrade their skills and keep abreast of the latest professional developments, as well as take advantage of opportunities for personal enrichment. Without these types of opportunities, it will be hard for people to navigate the transition to a digital economy. Investing in adult education will be a vital requirement as technology disruption accelerates.

PART III

AN ACTION PLAN

SEVEN

IS POLITICS UP TO THE TASK?

A MAJOR CHALLENGE in the current environment is how to generate a societal consensus around needed workforce and policy changes. If recent trends continue, it is possible that digital technologies and emerging business models will threaten existing practices of income provision, health benefits provision, and retirement support. Developed countries may end up with significant proportions of the population underemployed or unemployed, and that will pose risks to civil peace and prosperity.

There are many examples of developing nations around the world where the labor supply outpaces the number of jobs. Some countries in the Middle East, Africa, Latin America, and Southeast Asia are experiencing high economic inequality and a 30 to 40 percent youth unemployment rate. Large numbers of young people with little hope of bettering themselves can fuel social unrest and discontent with the existing regime. Some governments have resorted to severe repressive measures to maintain peace and keep disenchanted people from upsetting the public order.

An extreme version of this dystopia is shockingly described in the *Hunger Games* trilogy. It outlines a world of deprivation in the contrast between a wealthy capital city and a dozen poor districts in the hinterlands where people routinely starve to death. To entertain the masses and remind viewers of the power of the central city, leaders hold an annual competition in which a young boy and girl from each of the districts compete on television in a battle to the death. The losers die, the winner earns fame and glory, and the winner's hometown receives extra food and supplies.[1]

Despite looming threats in the developed world as technology disrupts the workforce, it is difficult in a polarized environment to get national leaders or the general public to think about digital disruption and the future of work. One exception is former president Barack Obama, who argued that "because of automation, because of globalization, we're going to have to examine the social compact, the same way we did early in the 19th century and then again during and after the Great Depression. The notion of a 40-hour workweek, a minimum wage, child labor laws, etc.—those will have to be updated for these new realities. But if we're smart right now, then we will build ourselves a runway to make that transition less abrupt."[2]

In this chapter, I discuss the challenges facing business and government leaders in how they construct a runway for workforce transitions. With technological innovation, changes in business models, and political discontent accelerating, there is a grave need to devise a new kind of politics that deals with economic dislocation and political dissatisfaction. Failure to address basic governance challenges and problems in the political system is a recipe for widespread unrest.

PAST EFFORTS TO ADDRESS MEGACHANGE

The history of civilization shows many bouts of large-scale transformation.[3] For example, the shift from an agrarian to an industrial economy was traumatic for workers. The historian Gregory Clark has estimated that real wages in the United Kingdom fell 10 percent between 1770 and 1810 and that real wage gains for English workers did not rise until sixty to seventy years after the onset of industrialization.[4]

Industrialization in the United States in the late nineteenth and early twentieth centuries also brought severe transition costs. The nation faced major challenges as new business models were put in place, characterized by the rise of mass production factories. These challenges included the need to retrain workers, address food safety problems, enact child labor laws, reduce economic concentration that was in the hands of a few, and manage the mass migration from the South to the large midwestern cities where factories were located. Many people did not have the skills needed to thrive in an industrial economy. They had grown up on farms, and moving to the city involved difficult cultural and economic adjustments.

It took a number of decades to resolve these tensions. There were violent work stoppages as employees unionized. Corruption was rampant in national, state, and local government. Maintaining food quality and achieving worker safety were key challenges when the market was largely unregulated. There was considerable turmoil throughout American society.

Yet through the far-sighted leadership of Presidents Theodore Roosevelt and Franklin Roosevelt and Secretary of Labor Frances Perkins, among others, the country devised new policies and built new business models. A combination of economic and political reforms helped government and business adapt to new conditions. Public policies were put in place to promote worker safety, improve food quality, and limit child labor. Large economic corporations were broken up to promote market competition. Novel political mechanisms such as primary elections, the direct election of senators, and state initiatives allowed ordinary people to participate in politics and choose their leaders. Social security and unemployment insurance programs were developed to help workers make the transition to an industrial era. Constitutional amendments were added to give women the right to vote and Congress the power to levy an income tax.

That period of change was not unique. Following World War II, European economies were devastated and the international order highly disrupted. The large combat death toll was shocking for many people as the war led to a major disruption of commerce, national security, and international relations. Yet new institutions such as the World Bank and the International Monetary Fund were created to facilitate foreign assistance and governance reform. Within a few years the United States had enacted the Marshall Plan and other multilateral aid programs that rebuilt Germany, France, and Japan and put them on the road to recovery. The world ended up stronger and more stable than before the war.

The post–World War II period represents an illuminating example of how the country and world came together to solve large-scale problems. Unlike the Industrial Revolution, which took decades of adjustment, the post–World War II era saw national and global leaders move quickly in just a few short years to make major decisions about political and economic issues. The results were positive for the globe and demonstrated that under the right conditions, leaders can cope with broad structural changes and improve the lives of many people.

DEALING WITH STRUCTURAL CHANGE

Despite the urgency of contemporary economic problems, it is hard for political leaders today to overcome polarization, hyperpolarization, and gridlock. Far-reaching economic transitions from the industrial to the digital world are hard to manage. Even though the need for new models of work and social service delivery systems is apparent, political and business leaders have difficulty coming up with realistic ideas the majority in Congress can agree on for income sustenance and benefit delivery.

This problem is aggravated by a lack of trust across partisan and ideological lines that makes it impossible to identify areas of possible agreement.[5] Compromise, bargaining, and negotiation have become anathema to politicians. The news media do little to help people understand what is at stake in the current challenges.

Some observers worry that American politics is moving toward a "post-truth" world in which facts don't matter and deceit and manipulation are commonplace. Former Federal Communications Commission chairperson Tom Wheeler writes that "technology has also become a tool to undermine truth and trust. The glue that holds institutions and governments together has been thinned and weakened by the unrestrained capabilities of technology exploited for commercial gain. The result has been to de-democratize the internet."[6]

Many aspects of politics, institutions, and technology make it challenging to devise alternative policies. The political system is fragmented and polarized. America's system of federalism and the separation of powers makes it difficult to solve problems. Rather than take thoughtful actions that would ameliorate pending problems, it is easier to pretend that nothing fundamental is happening and the need for action is not urgent.

Recent voting outcomes, such as the Brexit choice in the United Kingdom, the victory of Donald Trump in the U.S. presidential election, and the rise of ultra-nationalists in many places, signal public discontent on a broad scale. People argue over the role of trade and globalization and the appropriate size of government. Amid all the economic and political discontent, there is widespread disagreement over the scope of the problem and possible remedies for workforce development. It is important that these political problems do not distract attention from the economic disruption that is occurring and cloud efforts to address it.

THE TIE TO INEQUALITY

Inequality is a societal issue that makes economic dislocation and political discontent more difficult to address. Inequality conditions the overall environment in which policy discussions take place. The inequitable distribution of income and wealth makes it impossible to finance needed solutions or develop a social consensus on what needs to be done.[7]

The economists Thomas Piketty and Emmanuel Saez have documented the rise in income concentration over the past century. They chart the share of pre-tax income accrued to the top 1 percent of earners from 1913 to 2012.[8] In 1928, the year before the Great Depression, the top 1 percent garnered 21.1 percent of all income in the United States. Over the next fifty years, that figure dropped to a low of 8.3 percent in 1976, then rose to 21.5 percent in 2007. It dropped to 18.8 percent in 2011 after the global recession, then rose again to 19.6 percent in 2012.[9] These figures show that income concentration today is similar to what it was in the 1920s and more than double what it was in the post–World War II period.

More detailed statistics demonstrate that after-tax income stagnated for most workers from 1979 to 2009 while rising dramatically for the top 1 percent. Calculating the percentage change in real after-tax income for four groups of workers shows that during those thirty years, income rose 155 percent for the top 1 percent of earners, 58 percent for the next 19 percent of earners, 45 percent for the middle 60 percent, and 37 percent for the bottom 20 percent.[10] These patterns contribute to the public's disillusionment and feeling that the current system is rigged against ordinary people.

If Piketty's book, *Capital in the Twenty-First Century*, is correct, money is likely to become even more concentrated in the future. Drawing on data from several countries over the past 200 years, he argues that the appreciation of capital outpaces that of the economy at large and of wages in particular. That benefits the people who already hold a lot of financial resources and increases the overall concentration of wealth.[11]

Societal problems tied to inequality are very much connected with emerging technologies. Digital platforms have created tremendous wealth. Indeed, most of the large fortunes created by those under the age of forty have involved digital technology. Moreover, with innovation accelerating, the money tied to technology is likely to make inequality even more problematic in the future. As described by Colin Bradford, "The global economy appears to benefit the few rather than the many." He argues that though "technological change has increased the productivity of labor, labor has not received the incremental benefits of its own productivity improvement."[12]

Economic inequality is not just a financial challenge, it also affects politics and basic governance itself. The wealthy are much more politically active than the general public. In a "first-ever" public policy survey funded by the Russell Sage Foundation of "economically successful Americans," the political scientists Benjamin Page, Larry

Bartels, and Jason Seawright measured the activism and beliefs of the rich. They worked with the Wealthfinder "rank A" list of the top 2 percent of American households based on wealth and supplemented it with an Execureach list of high-level business executives of major companies. To reach their intended population, they screened for the top "1% of wealth-holders" and completed interviews with those individuals.[13]

In talking with them, the researchers found that 99 percent of the wealthy said that they voted in presidential elections, almost double the rate of the general public. Most (84 percent) also reported paying close attention to politics. Two-thirds (68 percent) made campaign contributions to politicians; in stark contrast, only 14 percent of the general public do.[14] Individuals with money, in general, are much more active politically than the general public.

The reason is clear: wealthy people know that political engagement matters.[15] Being involved in politics yields benefits and enables them to express their views and influence results. Unlike the general public, which tends to be cynical about politics, believing that there is no difference between Republicans and Democrats and that politics is not a very good way to produce change, many affluent people believe that politics matters and represents a way to affect national and international affairs. Indeed, a study by the political scientist Lee Drutman of the top 1,000 campaign donors from 2012 (those who gave at least $134,000) found that two-thirds favored Republicans and the largest number of them came from the financial sector.[16]

In light of the importance of engagement, it is not surprising that the ultra-rich report a large number of "high-level political contacts." When asked whether they had contacted public officials or their staffs in the preceding six months, 40 percent indicated they had contacted a U.S. senator, 37 percent had contacted a U.S. House member, 21 percent had contacted a regulatory official, 14 percent had contacted someone in the executive branch, and 12 percent had contacted a White House official.[17] Those rates are much higher than the rates for the general public. A national survey undertaken at the University of Michigan documented that about 20 percent of ordinary people said they had contacted a member of the U.S. Senate or House in the preceding four years, through telephone calls, letters, or visits to legislative offices.[18]

Political activism matters because the super-rich, as a group, hold policy views that are significantly different from the views held by ordinary citizens. In their survey, Page, Bartels, and Seawright asked the wealthy about a range of public policy issues.[19] Comparing their opinions with those of the general public, the researchers found that top wealth-holders "differ rather sharply from the American public on a number of important policies. For example, there are significant differences on issues such as taxation, economic regulation, and social welfare programs."[20] Research summarizes the gulf in policy preferences between the top 1 percent and the general public. The wealthy are more likely than the general public to favor cuts in Medicare and education (58 percent versus 27 percent for the public) and less likely than the general public to believe the government has an essential role to play in regulating the market (55 percent versus 71 percent, respectively).

Most surprising, however, are the differences in views about social opportunities. In the abstract, it might be assumed that there would be little gap in this area. According to the Credit Suisse Global Wealth Database, two-thirds (69 percent) of wealthy individuals come from humble origins and conceivably could favor a limited government role in the economy but still value equity of opportunity.[21] But that is not what Page, Bartels, and Seawright found in their survey. Their data show that while 87 percent of the general public believe the government should spend whatever is necessary to ensure that all children have good public schools, only 35 percent of the top 1 percent do.[22] The wealthy also are less likely than the general public to want the government to provide jobs if private sector positions are unavailable, to believe the government should provide a decent standard of living for the unemployed, or to be willing to pay more taxes to support universal health care.

This research indicates that those with great resources are far more conservative than the general public on a range of issues related to social opportunity, education, and health care. They do not support a major role for the public sector, even when government actions would further economic and social opportunities for the public. They are much more likely to favor cuts in social benefits and programs that benefit less fortunate members of society. These views of the superrich lead them to favor tax cuts, even though tax cuts reduce the financial resources available to invest in education and health care. If politically active rich people favor tax cuts and support austerity measures, as has been the case in recent years, it is difficult to generate support for programs that help the nation's low- and middle-income people improve their lot.

According to the Princeton sociologist Martin Gilens, there is a strong link between "affluence and influence." Through a detailed analysis of policymaking, public opinion, and income levels, he demonstrates that "affluent Americans' preferences exhibit a substantial [positive] relationship with policy outcomes whether their preferences are shared by lower-income groups or not." Gilens argues that there is "virtually no relationship between policy outcomes and the desires of less advantaged groups" when the preferences of the latter diverge from those of the wealthy.[23] And in a follow-up study with Benjamin Page, he examined the impact of average citizens and economic elites on 1,779 different policy issues over the past thirty years and concluded that ordinary people had "little or no independent influence."[24]

The substantial connection between wealth and political influence makes ordinary citizens very cynical about the political system. When they see wealthy interests exercising disproportionate influence and gaining undue benefits, they conclude that the system is rigged against them and that basic governance is flawed as well. Those beliefs make it difficult to build support for government action, even for programs designed to help ordinary individuals improve their economic lot.

THE NEED FOR FLEXICURITY

If countries end up in a situation in which many people are unemployed or underemployed for significant periods of time, their workers will need some way to receive health care, disability, and pension benefits outside of employment. Called "flexicurity," or flexible security, this idea "separate[s] the provision of benefits from jobs."[25] According to Jean Pisani-Ferry, a labor adviser to French president Emmanuel Macron, it is important to support workers in general rather than the preservation of specific jobs. "The reforms currently under discussion combine a broadening of the access to unemployment insurance that would eventually turn it into a universal safety net for all those suffering an income drop as a consequence of economic disruption," he notes.[26]

Currently, in the United States, when they are fully employed, people are eligible for company-sponsored health care plans and pensions. That approach functioned well in an era when most Americans who wanted jobs were able to get them. People with limited skills were able to find well-paying jobs with benefits in factories, warehouses, and production facilities. They could educate their children, achieve a reasonable standard of living, and guard against disabling illnesses.

The complication came when the economy changed over the last couple of decades, wages stagnated, and technology made it possible for companies to get by with fewer full-time workers. In conjunction with the introduction of robotics and AI into the workplace, jobs have disappeared in certain sectors and businesses have shifted to temporary or offshore employees.

Some countries have experimented with a short workweek. For example, Gothenburg, Sweden, undertook a two-year trial project in which retirement home workers went from an eight-hour workday to a six-hour workday at the same level of pay. The adjustment necessitated the hiring of seventeen new nurses at a cost of $738,000 a year but resulted in "happier, healthier and more productive employees."[27] That success has encouraged other communities to implement a similar change.

France has pioneered such an approach on a national level. Fifteen years ago, it moved to a thirty-five-hour workweek. This change, however, did not make a dent in the country's historically high unemployment level. The nation still has one of the highest rates in Europe as over 10 percent of its residents have been unable to find gainful employment.

The need to come up with creative models will challenge existing political patterns and necessitate different ways of thinking about governance. In the same way that the early twentieth century was tumultuous owing to the transition from an agrarian to an industrial economy, the twenty-first century will be a time of upheaval as each country grapples with ways to help people confronting deep structural change. These challenges will vex leaders in many countries as they experience the backlash from economic dislocation.

One example is the way some emerging technologies lower government revenues. Many local units currently rely on parking fees and fines for moving violations to fill out their budgets. As autonomous vehicles become widespread, local governments likely will lose money from traffic violations because these cars are not likely to be speeding, going through stop signs, having accidents, or committing moving violations. According to Kevin Desouza of Arizona State University, "New [urban] innovations as simple as phone apps in combination with routine improvements like meter upkeep have already accidentally reduced parking ticket revenue [in Washington, D.C.], with a drop from $90,610,266 in 2012 to $84,458,255 in 2013." He pointed out that "users can use their smartphones to remotely feed their meters before they expire [and] submit parking ticket photos and enter violation codes to an app that provides helpful information on getting the ticket dismissed."[28]

A major challenge going forward is to use technology to bring the benefits of the digital revolution to a wider range of people. Right now, 20 percent of Americans and nearly half of the world's population lack digital access. An estimated 4 billion people around the globe do not have internet access. This limits the impact of the technological revolution and prevents those lacking access from gaining the benefits that are there. Unless we can bring more people into the digital era, it will be impossible to address fundamental inequalities derived from technological innovation.

THE RISKS OF INACTION

The United States is in the early stages of its transition to a digital economy. Developments such as robotics and AI are unfolding, and it will take a while for these novelties to alter the workforce and have a dramatic effect on people's economic lives. It may be a decade or two before the full force of these changes is felt by a substantial number of people.

As these trends intensify, though, large-scale solutions and fundamental policy remedies will be needed. Inaction early in the transition worsens inequality and increases social and economic tensions down the road. If new policies were adopted now, they could go a long way toward easing the impact of societal transformations.

As noted by former Federal Reserve Bank chairman Ben Bernanke, "Economic growth is a good thing." However, recent political developments have brought home the idea that "growth is not always enough."[29] By itself, growth does not reduce inequality, even if it helps some people weather the financial storm. Sometimes major developments require strong and forceful responses to deal with underlying problems and address fundamental dislocations.

The economist Raj Chetty has analyzed millions of tax records over several generations of Americans and found that only 50 percent of people born during the 1980s will earn more than their parents, compared to 92 percent of those born in the 1940s, 79 percent of those born in the 1950s, 62 percent of people born in the 1960s, and 61 percent of individuals born in the 1970s.[30] This decrease in economic mobility, fueled by the rising level of inequality in America, frustrates many people and weakens overall economic prosperity.

The inability of the U.S. economic and political system to address basic problems angers ordinary citizens and intensifies mistrust in democratic institutions. People no longer believe that political leaders can improve their lives. They doubt the accuracy or fairness of news coverage. These kinds of frustrations lead to populist uprisings and the potential for widespread unhappiness. Governance failures, real or imagined, make it difficult to resolve problems generated by structural economic change. Dealing with these issues now before the ramifications exceed policymakers' grasp will help society in the long run.

WHY TRUMPISM IS NOT A POLITICAL ABERRATION

In the current turmoil, politicians debate the causes of and remedies for economic discontent. As an example, President Donald Trump has defined the economic problem in particular ways. He focuses on manufacturing jobs and trade agreements as the primary source of economic dislocation and weak financial prospects. He believes that these issues are responsible for the poor plight of the working class and that action on these fronts will broaden prosperity and help more people cope with these economic changes.

However, his diagnosis is too limited in how it defines the scope of economic difficulties. He does not understand that disruption is no longer limited to the manufacturing sector but is spreading across a number of economic areas. The combination of technological change, changes in business models, and the emergence of the sharing economy is transforming the workforce and creating broad fiscal challenges.

Trump himself may be a transitional political leader, but the cleavages he identified are real and enduring. Structural change is not likely to go away when he leaves office. If anything, the social, economic, and political tensions he has underscored are likely to advance in scope and intensity. Recent populist uprisings such as the tea party, anti-globalization sentiments, anti-Wall Street viewpoints, and the Trump candidacy itself may look mild compared to what could emerge in the future.[31]

This is why early action is needed before resentment, anger, and unrest reach hurricane levels. It is certainly the case that a number of workers will see their jobs affected or even eliminated, which will lead to considerable anger and anxiety over the pace of economic change. Those of average means are likely to feel powerless in the face of broad-scale transformation, and this will aggravate public anger directed toward establishment elites.

There already is solid evidence of a link between manufacturing job losses and recent voting patterns. A county-level analysis of the 2016 presidential election, for example, found that "most of the zones where more than one robot was introduced for every thousand workers ended up backing Trump over Clinton."[32] There clearly was an association between places that introduced robots, factories that lost manufacturing jobs, and votes for the Republican presidential candidate.

The discontent that has arisen in recent years in advanced economies may be just a small-scale precursor of what could emerge in following decades. Trump, for example, correctly articulated that people of average means are at great risk in the current situation. A number of them already have been hurt by the loss of manufacturing jobs. As autonomous vehicles replace trucking jobs, mobile tablets replace restaurant wait staffs, and purchasing apps replace sales clerks, the impact on middle America could become far more pronounced.

In today's complex and chaotic world, people are worried about economic trends and looking for new identities and a sense of purpose. As explained by Facebook chief executive officer Mark Zuckerberg in a Harvard University address, "Purpose is that sense that we are part of something bigger than ourselves, that we are needed, that we have something better ahead to work for. Purpose is what creates true happiness. You're graduating at a time when this is especially important. When our parents graduated, purpose reliably came from your job, your church, your community. But today, technology and automation are eliminating many jobs. Membership in communities is declining. Many people feel disconnected and depressed, and are trying to fill a void."[33]

THE MISMATCH BETWEEN ECONOMIC OUTPUT AND POLITICAL REPRESENTATION

Part of the looming governance crisis in the United States is the stark mismatch between economic output and political representation. A 2016 Brookings Institution study by Mark Muro and Sifan Liu found that only 15 percent of the counties in America generated 64 percent of gross domestic product.[34] These prosperous areas were mostly urban areas on the East and West Coasts, with a few scattered metropolitan areas in between, and these places largely voted in favor of the Democratic Party candidate Hillary Clinton in the 2016 presidential campaign. In contrast, the remaining 85 percent of the counties voted for Trump and generated only 36 percent of GDP. Those were areas where people felt that prosperity had bypassed them and the system was rigged against those of ordinary means.

These results were echoed in a study by the Economic Innovation Group. It explored the geographic basis of economic development and found "U.S. geographical economic inequality is growing" and "a large portion of the country is being left behind by today's economy." This organization analyzed job creation at the county level and found that "new jobs are clustered in the economy's best-off places" and that only 25 percent of new jobs are being created in the poorest 60 percent of counties.[35]

Part of the problem has been a slowdown in the startup economy in the aftermath of the global recession. Small businesses typically generate a substantial proportion of new jobs, although sometimes these are temporary or part-time positions with limited benefits. For example, 414,000 firms were launched in 2015, down from 558,000 in 2006.[36] In the digital era, it is hard for small companies to compete with large firms having network advantages and a much larger scale.[37] As noted by the University of Montreal professor Yoshua Bengio, a pioneer in the AI field, there are dramatic advantages to scale. "More data and a larger customer base gives you an advantage that is hard to dislodge. Scientists want to go to the best places. The company with the best research labs will attract the best talent. It becomes a concentration of wealth and power," he writes.[38]

In addition, economic development is complicated by significant geographic disparities among startups. Most of them are launched on the East or West Coast, which furthers the mismatch in economic activity across the regions. The reason is clear: 75 percent of venture capital dollars go to firms in California, New York, and Massachusetts.[39] That fuels economic activity on the coasts, not in the hinterlands.

Similar patterns have developed internationally. Research by Richard Florida has found that "the fifty largest metropolitan areas house just 7 percent of the world's population but generate 40 percent of its growth." From his standpoint, these "superstar" cities have become "gated communities" that promote inequality and destroy the vibrancy of urban areas.[40]

Politically, this economic divergence is devastating because there is little reason to believe it will decrease any time soon. If anything, the disparity between economic output and political representation is likely to grow. As automation, robotics, and AI grow in the economy in the most prosperous areas, these developments could actually increase inequity and deepen the crisis of democracy.

Reflecting on this situation, Brookings scholar Muro has pointed out that "this is a picture of a very polarized and increasingly concentrated economy, with the Democratic base aligning more to the more concentrated modern economy, but a lot of votes and anger to be had in the rest of the country."[41] The political consequences of this emerging pattern could be quite substantial. As workers in the less prosperous areas fall further behind, they likely will grow angrier and push aggressively against globalization, free trade, immigration, and open economies.

Similar to the Trump agenda, they will want to decrease labor trends that they think threaten their jobs, slow the innovation they believe is limiting their employment opportunities, and oppose government spending on programs that help workers adjust to a digital economy. As Muro presciently points out, "We're going to have a lot of questions about how to translate the political geography into actually helpful policy."[42]

As leaders grapple with these issues, there could be damaging consequences for the political system and the ability to enact forward-looking legislation. Representation in the U.S. Senate, for example, is based on two senators per state, regardless of a state's population or economic activity. The result of current economic patterns in this governing context is that the economically stagnant areas could end up electing two-thirds of U.S. senators while the prosperous areas could elect only one-third.

This is a formula for political disaster. There cannot be that wide a disparity between political and economic forces without severe repercussions. Over the coming years, as the technology revolution unfolds and generates considerable economic dislocation, the ingredients are in place for a full-blown political backlash from voters who are angry that their geographic areas are stagnating and not sharing in the prosperity of the two coasts. Because of the institutional framework of U.S. governance, they will be in a strong position to express their anger and block action they believe ill-advised or too costly for taxpayers. Unless the discontent associated with serious economic disparities is addressed, public unhappiness could prevent needed policy responses and exacerbate existing social and economic tensions in the United States.

MEDIA CHAOS AND DISINFORMATION

The news media landscape is in a state of considerable flux and poorly designed to face the difficulties raised by economic disruption and technological innovation. At the same time, disinformation and hoaxes, popularly referred to as "fake news," are increasingly affecting the way individuals interpret daily developments. Information systems have become more polarized and contentious, and there has been a precipitous decline in public trust in traditional journalism.[43]

As the overall media landscape has changed, several ominous developments have taken shape. Rather than using digital tools to inform people and elevate civic discussion, some individuals and groups have taken advantage of social and digital platforms to deceive, mislead, or harm others through disseminating fake news and disinformation with automated bots or AI algorithms. False news and disinformation campaigns are being generated by outlets that masquerade as actual media sites but promulgate misleading accounts designed to deceive the public. When these efforts move from sporadic and haphazard to organized and systematic, they develop the potential to disrupt campaigns and governance in entire countries.[44]

As an illustration, the United States witnessed organized efforts to disseminate false material in the 2016 presidential election. A Buzzfeed analysis found that the most widely shared fake news stories in 2016 concerned "Pope Francis endorsing Donald Trump, Hillary Clinton selling weapons to ISIS, Hillary Clinton being disqualified from holding federal office, and the FBI director receiving millions from the Clinton Foundation."[45] Using a social media assessment, Buzzfeed claimed that the twenty largest fake stories generated 8.7 million shares, reactions, and comments, compared to 7.4 million generated by the top twenty (reputable) stories from nineteen major news sites.

Fake content was widespread during the 2016 presidential campaign. Facebook has estimated that 126 million of its platform users saw articles and posts promulgated by Russian sources. Twitter has found 2,752 accounts established by Russian groups that tweeted 1.4 million times in 2016.[46] The widespread nature of these disinformation efforts led Columbia Law School professor Tim Wu to ask, "Did Twitter kill the First Amendment?"[47]

A specific example of disinformation was the so-called Pizzagate conspiracy, which started on Twitter. The story falsely alleged that sexually abused children were hidden at Comet Ping Pong, a Washington, D.C., pizza parlor, and that Hillary Clinton knew about the sex ring. It seemed so realistic to some that a North Carolina man named Edgar Welch drove to the capital city with an assault weapon to personally search for the abused kids. After being arrested by the police, Welch said "that he had read online that the Comet restaurant was harboring child sex slaves and that he wanted to see for himself if they were there. [Welch] stated that he was armed."[48]

An analysis after the election found that automated bots played a major role in disseminating false information on Twitter. According to Jonathan Albright, an assistant professor of media analytics at Elon University, "What bots are doing is really getting this thing trending on Twitter. These bots are providing the online crowds that are providing legitimacy."[49] With digital content, the more posts that are shared or liked, the more traffic they generate. Through these means, it becomes relatively easy to spread fake information over the internet. For example, as graphic content spreads, often with inflammatory comments attached, it can go viral and be seen as credible information by people far from the original post.

A postelection survey of 3,015 American adults suggested that it is difficult for news consumers to distinguish fake from real news. Chris Jackson of Ipsos Public Affairs undertook a survey that found "fake news headlines fool American adults about 75 percent of the time" and "'fake news' was remembered by a significant portion of the electorate and those stories were seen as credible."[50] Another online survey of 1,200 individuals after the election by Hunt Allcott and Matthew Gentzkow found that half of those who saw these fake stories believed their content.[51]

Fake news stories are amplified and disseminated quickly through false accounts, or automated bots (short for robot). Most bots are benign in nature, and some major sites, such as Facebook, ban bots and seek to remove them, but certain social bots are "malicious entities designed specifically with the purpose to harm. These bots mislead, exploit, and manipulate social media discourse with rumors, spam, malware, misinformation, slander, or even just noise."[52]

This information can distort election campaigns, affect public perceptions, and shape human emotions—even drive electoral choices. Recent research has found that "elusive bots could easily infiltrate a population of unaware humans and manipulate them to affect their perception of reality, with unpredictable results."[53] In some cases, they can "engage in more complex types of interactions, such as entertaining conversations with other people, commenting on their posts, and answering their questions." Through designated keywords and interactions with influential posters, they can magnify their influence and affect national or global conversations, especially resonating with like-minded clusters of people.[54]

False information is dangerous because of its ability to affect public opinion and electoral discourse. According to David Lazer and colleagues, "Such situations can enable discriminatory and inflammatory ideas to enter public discourse and be treated as fact. Once embedded, such ideas can in turn be used to create scapegoats, to normalize prejudices, to harden us-versus-them mentalities and even, in extreme cases, to catalyze and justify violence."[55] As they point out, factors such as source credibility, repetition, and social pressure affect information flows and the extent to which misinformation is taken seriously. When viewers see trusted sources repeat certain points, they are more likely to be influenced by that material.

Recent polling data demonstrate how harmful these practices have become to the reputations of reputable platforms. According to the Reuters Institute for the Study of Journalism, only 24 percent of Americans today believe social media sites "do a good job separating fact from fiction, compared to 40 percent for the news media."[56] That demonstrates how much these developments have hurt public discourse.

CONCLUSION

Getting to yes on necessary reforms will take a major effort on the part of political and business leaders. Today's polarized rhetoric and problems in the media system do not bode well for the nation's ability to address these issues and help people deal with transition difficulties. It is challenging to address even minor questions of public policy, let alone divisive questions about inequality, income distribution, public policy, and attributions of responsibility for economic outcomes. News media coverage is superficial and uninformative.

How in this limited-information environment leaders will be able to address far-reaching and controversial economic issues remains an open question.[57]

There is no guarantee that the change from an industrial to a digital economy will go smoothly. At least in the short run, emerging technologies are likely to worsen inequality, widen the economic gap between the coasts and the hinterland, mislead people, and stoke social and political tensions. Societal conditions could deteriorate significantly or spiral completely out of control. Neither our political leaders nor voters in general are prepared to have the kinds of conversations that are needed right now. Much of the current divide is left versus right or Republican versus Democrat.[58]

Those partisan issues are mild in comparison to future discussions, which will center on the social responsibility for millions of people who want to work but are unable to do so because they are not needed in the workforce. The question of social versus personal responsibility for poor economic outcomes cuts squarely against the American tradition of personal freedom and individual attributions of responsibility. The emerging economy will vex existing political alliances as well as media coverage and system governability. It will test people's confidence in leaders and how they handle the coming digital transformation.

EIGHT
ECONOMIC AND POLITICAL REFORM

THERE IS A serious risk that political leaders will misdiagnose the current situation and make the workforce problem worse through shortsighted or ill-informed decisions. They could take steps that destabilize society, aggravate economic tensions, and advance authoritarian measures in order to maintain public order. Unwittingly, they could turn individual communities or nations as a whole into dystopias that rival the unhappy scenarios of science fiction movies.[1]

In the current situation, though, several economic reforms could ease the transition to a digital economy. These measures include developing a new concept of remunerated work to include parenting, mentoring, and volunteering; enacting paid family and medical leave; expanding the earned income tax credit; and improving health, education, and well-being.

At the same time, fundamental changes are needed in how the U.S. political system operates. Potential reforms in this sphere include building a Republic 2.0 politics that addresses economic dislocations, enacting universal voting to reduce political polarization, reducing geography-based inequities, improving legislative representation, abolishing the Electoral College, enacting campaign finance reform, and adopting a solidarity tax to finance needed social programs. Implementing these measures would help people cope better with the looming disruptions.

NEW MODELS OF WORK

Many workers derive their personal and professional identity at least partly from their jobs. Their position helps articulate their sense of purpose and contributes to their social network of friends and colleagues. The employment title provides income, benefits, and retirement security. Going to the office offers a structure to the day and a rhythm to the week. Paid vacation time is covered by the employer and takes place when it fits within the business's work schedule. Some companies even organize social outings, picnics, and affinity groups for employees to help foster a sense of camaraderie and teamwork.

Yet this job-focused construct is a recent notion in human life. For much of recorded history, jobs were not the be-all and end-all of human existence. People understood their identity as more closely linked to family, ethnic group, religion, neighborhood, or tribe. They worked enough to earn a subsistence existence, but their employment did not dominate their social existence. They did not spend most of their waking hours either working or thinking about work.

In the future, people may revert to this historical way of life. Their job (or, more likely, jobs) may be part of who they are, but will not constitute their total existence. As the economic dislocations associated with technological innovation accelerate, people's lives will become more multifaceted to reflect more of the variety of things that people can do and the multifarious activities that engage their interests. These include such activities as parenting, mentoring, and volunteering, and hobbies such as sports or the arts. People will have time to seek a better work-life balance and carve out time for pursuing personal interests.

As was true of many traditional societies, individuals contribute in different ways to their community, and having a broader conception of life will liberate people to pursue a number of different activities. Rather than being consumed by a job, they will see their position as only part of what they do. It will not dominate their waking hours or make up their total identity.

The transition to a digital world will necessitate the adoption of new policies related to work. The United States remains the only major developed country without an official program that pays people to take care of newborn infants or elderly parents. These are exactly the types of activities that are socially desirable and benefit the overall community. As people broaden their conception of jobs to deal with technological change, this initiative is one of the most crucial proposals to adopt.

Such an initiative would offer clear benefits in terms of health and longevity. Researchers have found that those who receive paid family or medical leave have healthier babies and better life outcomes.[2] The evidence pointing to positive outcomes associated with this kind of financial support is quite strong. In countries where such leave is provided, recipients appreciate it and make productive use of the time supporting their family.

During a period of economic and technological transition, this proposal is especially relevant. Emerging technologies and new business models facilitated by mobile technology suggest a need to redefine our conception of work to include tasks that are socially beneficial even if they are not currently remunerated. These are exactly the sorts of initiatives that would help people as they transition to new kinds of jobs.

For this conception of work to be viable, though, there must be reasonable ways for ordinary people to generate income and social benefits outside the provisions of a full-time job. Individuals cannot engage in such socially beneficial activities as parenting, caregiving, or mentoring unless social and economic value is attached to those efforts. Community-benefiting activities should be important enough to be considered worthy of health and retirement benefits.

Earlier chapters detailed several ways in which expanded social benefits could be implemented. Means include privately operated citizen accounts, worker-controlled benefits, or government-run benefit exchanges.[3] Depending on one's views about the proper role of government and the importance of private or nonprofit organizations, people may lean toward a public, private, or nonprofit variant of benefit provision. It matters less how this service is organized than the principles that workers should control their benefits and that benefits should be portable across sectors and geographically.

Benefit portability is a key attribute in the current era because it helps individuals navigate a world of part-time jobs or volunteer commitments that do not provide benefits. These are important considerations since they are the types of work situation likely to increase in coming years. As economic disruption unfolds and people are affected by structural transformation, a number of individuals will end up with something other than a full-time job with paid benefits. How we handle such workers and their incomes and benefits will dictate the kind of world in which we live.

IMPROVING HEALTH, EDUCATION, AND WELL-BEING

Inequality is not just an economic issue, it has major consequences for health and well-being. For example, studies have found an association between economic inequality and mortality rates. An analysis by the Princeton University economists Anne Case and Angus Deaton shows in recent years "increases in drug overdoses, suicide, and alcohol-related mortality, particularly among those with a high-school degree or less."[4] These are exactly the people who have fared poorly in recent decades and whose jobs are at risk from automation. As their economic situation has worsened, these individuals have faced more ill health and limited longevity. As Case and Deaton write, "Cumulative disadvantage . . . in the labor market, in marriage and child outcomes, and in health, is triggered by progressively worsening labor market opportunities."

Brookings Institution scholars Carol Graham, Sergio Pinto, and John Juneau find similar results. These researchers compared personal well-being sentiments to mortality rates and found "a robust association between lack of hope (and high levels of worry) among poor whites and the premature mortality rates."[5] They tie what they call the "geography of desperation" to technological changes affecting the workforce, noting that "a critical factor is the plight of the white blue-collar worker, for whom hopes for making it to a stable, middle-class life have largely disappeared. Due in large part to technology-driven growth, blue-collar jobs in the traditional primary and secondary industries—such as coal mining and car factories—are gradually disappearing."[6]

The demise of the American Dream and the social mobility associated with it has harmed the happiness and personal well-being of those in the middle and at the bottom of the economic scale. It has created personal problems for those affected by technological change and has challenged the fortunes of society as a whole. Not only do we need to ameliorate the deleterious ramifications of economic inequality, there must be programs to help those whose health and education have suffered from the combination of technological innovation and economic disruption.

One option that has yielded favorable results is preschool programs. Considerable research supports the health and learning benefits of enrichment classes for three- and four-year-olds. Those children who participate in such activities usually experience an "early boost," especially underserved children. A detailed analysis of 6,150 children from birth to five years old revealed that "the nation's average 4-year-old attending a typical pre-K does show slightly accelerated cognitive growth when compared with peers who remain in their home with a parent or informal caregiver."[7]

Improving worker training is another key objective. In general, community colleges offer options that are low cost and accessible to workers. Many businesses work closely with these and other educational institutions to improve workers' skills.[8] They draw on retraining programs to upgrade their workforce and provide new opportunities for employees. New advances in distance learning also provide opportunities for adults to upgrade their skills if they are outside the workforce. Making sure retraining programs position the individual for success on an enduring basis is important to the ultimate success of those efforts.

Finally, it is important to address the epidemic of opioid abuse that plagues working-class and professional communities. This issue deserves special attention during a time of economic transformation because the psychological stresses unleashed by structural changes have increased substance abuse and the mental health issues associated with financial setbacks. Rural areas, small towns, and urban communities all have reported thousands of drug overdoses in recent years. When people face great stress in their personal lives, it is tempting to turn to painkillers, tranquilizers, methamphetamines, cocaine, or heroin.

When overdoses result, drugs such as naloxone can revive the victims. Nevertheless, it costs $4,500 for injections and $150 for nasal inhalers, and this added cost taxes the budgets of communities across America. Even though naloxone can save lives, some cities have stopped stocking this medication for budgetary reasons. They refuse to allow their law enforcement officers or emergency personnel to administer the antidote even though its medical value is clearly apparent.[9]

Congress has authorized around $1.1 billion to deal with opioid abuse. But that is a drop in the bucket in light of the magnitude of the drug abuse problem. The number of people affected is quite large, and their mental health problems are severe. Substantial resources will be needed to combat this scourge. Moreover, as technological innovation expands and threatens people's livelihoods, the scope of this health concern likely will expand dramatically and create an even larger social problem.

REPUBLIC 2.0: THE NEED FOR POLITICAL REFORM

Economic change is not the only thing required in the current period. To make progress on the various initiatives discussed in this book, a new kind of politics is needed, one that allows more substantive policy discussions and a greater capacity to make effective decisions. In a situation where "real median household income grew at 0.5% per year from 1984 to 2015," the contemporary dialogue focuses heavily on worn-out clichés regarding ways to stimulate the economy and the proper role of government in economic development.[10] Conservatives want less government, while liberals believe the public sector performs important tasks in stabilizing society, dealing with imperfect markets, and helping people adjust to the future. Anger and disillusionment play prominent roles as each side tries to gain an advantage with a highly polarized electorate.

The current stalemate has led the economists Daron Acemoglu and Simon Johnson to propose a fundamental rethinking of political institutions along the lines of the Progressive movement that restructured politics at the turn of the twentieth century. They pinpoint actions such as the Sherman Antitrust Act of 1890, the 1913 income tax (adopted through a constitutional amendment), provision for the direct election of U.S. senators in 1913, an amendment guaranteeing women the right to vote in 1920, and state-level adoption of primaries and referendums as key actions that broke the logjam of that era. They write, "The progressives understood, through bitter experience, that politics could not be separated from economics. Their response to the economic ills created by the concentrated power of business was to demand not just economic reform, but political reform—on critical issues going so far as to amend the Constitution."[11]

In their book, *Democracy in America*, the political scientists Benjamin Page and Martin Gilens echo these sentiments. They argue that U.S. democracy has "gone wrong" through the unequal wealth that has distorted politics, the political clout of wealthy individuals, and the hijacking of representative government by corporations and interest groups. What is needed, in their view, is a "social movement for democracy" that democratizes institutions and provides a more equal voice for ordinary citizens.[12]

Brookings scholars William Galston and Clara Hendrickson find merit in tougher antitrust enforcement. Looking at the growing concentration of numerous economic sectors, they say it is time for four actions: tighter enforcement of horizontal mergers, updating the guidelines for nonhorizontal mergers, cracking down on predatory pricing, and reducing the costs of antitrust enforcement. If startups are to have a fair shot at competing in the marketplace, they will need policies that promote competition.[13]

In looking at the contemporary situation, unfortunately, there are several political and institutional barriers that prevent the adoption of needed reforms. The current climate is overly polarized and completely dysfunctional. Fundamental disagreements make it impossible for leaders to address the impact of technological change. They simply cannot get to yes on adopting needed actions.

Communications reform is needed to break this logjam. Former Federal Communications Commission chairperson Tom Wheeler advocates "public interest APIs" (application programming interfaces) that provide smart network oversight of social media platforms. He thinks that algorithms could help readers and viewers discern when they are being subjected to online manipulation, deception, or deceit. His hope is that digital tools can protect democratic norms "without intrusive government micro-management or bureaucracy."[14]

In addition, Clara Hendrickson and William Galston argue that "tackling automation would require modernizing the social safety net to not only improve the working lives of Americans today, but also to respond to the changing nature of work itself. . . . This might require openness to challenging some of the premises that have long guided partisan economic platforms." According to their perspective, "traditional conservatives will have to embrace some degree of government intervention."[15]

Furthermore, people will have to reconsider their views about social responsibility. Currently there is a major divide between those who believe people should be responsible for themselves and those who think there is a social responsibility for everyone in the community. Those emphasizing personal responsibility argue that if someone does well, that person should reap the rewards of her or his work and ideas, not pay too much in taxes, and not be subject to income redistribution. Moreover, if someone does not do very well, it is likely that person's own fault for not working hard enough, not having the right skills, or not being highly motivated.

In the future, Americans will need a responsibility framework in which individuals help others if they wish to avoid widespread social unrest. There could be many hard-working and highly motivated workers who are unable to find jobs. It is not that they are lazy, undeserving, or unwilling to work. Rather, many individuals may not be needed in the workforce as automation, robotics, and AI reach the point that fewer employees are needed to produce the necessary output.

Other countries with a solidarity culture will have an easier time dealing with workforce changes. They have societal norms that support taking care of less fortunate residents and helping them get back on their feet. They don't have the individualistic perspective that perpetuates the view that people are responsible for themselves and no one else. Many of these societies already have government programs based on social responsibility that can be expanded if unemployment spreads.

It will be exceedingly difficult for the United States to move toward a societal consensus on workforce issues because the contemporary situation is plagued by government gridlock, widespread public mistrust of leadership, and ill will between people of different persuasions. It is impossible to get people to accept the same facts, let alone the importance of considering alternative policy initiatives. In a situation where respected analysts believe a significant percentage of individuals face unemployment or underemployment, leaders in both parties will have to figure out ways to build consensus or at least majority support for needed actions.[16]

The writers E J. Dionne, Norm Ornstein, and Tom Mann complain about the rigged nature of contemporary politics. They note that "our system is now biased against the American majority because of partisan redistricting (which distorts the outcome of legislative elections), the nature of representation in the United States Senate (which vastly underrepresents residents of larger states), the growing role of money in politics (which empowers a very small economic elite), the workings of the Electoral College (which is increasingly out of sync with the distribution of our population) and the ability of legislatures to use a variety of measures, from voter ID laws to the disenfranchisement of former felons, to obstruct the path of millions of Americans to the ballot box."[17]

During an adjustment period, which may last for several decades, government likely will play a bigger role in society because private markets alone will not be able to address the economic disruption caused by structural transformations. Similar to what happened with the shift to an industrial economy at the turn of the twentieth century, new programs will be needed to help workers cope with fundamental changes and enable them to develop new skills.

These policy moves will be challenging because Americans have complicated views about individual freedom and the proper role of government. It will be hard to redesign the social contract, develop new programs, and identify new revenue sources if a sizable number of people continue to favor personal attributions of responsibility. Leaders have to do a better job explaining why they think their proposals will benefit average workers. This is not a strong suit of our current political discourse, so getting people to agree on any initiative will be a major barrier to handling the coming turmoil. As was the case in the early twentieth century, it will take decades to work through these issues.

ENACTING UNIVERSAL VOTING TO REDUCE POLITICAL POLARIZATION

One of the biggest problems to be faced in addressing the effects of structural economic transformation is political polarization. In the United States, low voter turnout has encouraged a toxic brand of extremist politics and party polarization.[18] When only 55 to 60 percent of eligible voters cast ballots in presidential elections, around 40 percent vote in off-year congressional contests, and only 10 to 20 percent in local elections, politicians have incentives to play to the base, take extreme positions, and eschew bargaining and negotiation as signs of a lack of political principle. Rather than focusing on the middle and seeking compromise, too often they take nonnegotiable stances and refuse to acknowledge the validity of alternative viewpoints. The result is a system that is gridlocked and plagued by a winner-take-all mentality.

Yet there is an idea that could help dramatically with polarization. That is universal voting, which offers the potential to disrupt the dysfunctional aspects of contemporary politics.[19] More than two dozen countries, including Australia and Belgium, require citizens to vote; nonvoters pay a small civil fine if they fail to comply.[20] Nearly all the countries with this approach have stable politics and electoral participation that is well over 90 percent.

This type of reform could fundamentally alter the electoral incentives facing leaders, creating the possibility for more broad-based appeals and less political extremism. According to Brookings Institution scholars William Galston and E. J. Dionne, "Intense partisans are more likely to participate in lower-turnout elections while those who are less ideologically committed and less fervent about specific issues are more likely to stay home."[21]

Having a universal voting requirement would change this paralyzing electoral dynamic and improve the functioning of democracies. It would alter the strategic environment facing candidates, decrease the incentives for extremism, and put our leaders in a better position to take pragmatic action. These are important considerations because the coming years will test the governability of our political system and the ability of leaders to offer bold and effective programs.

REDUCING GEOGRAPHIC INEQUITIES

Mismatches based on geography have created serious problems for American democracy. The two coasts and urban areas in general have fared much better in terms of economic prosperity than the rural hinterland. In addition, people with just a high school education or less have fared much worse than those with a college education.[22]

A public opinion survey undertaken by the *Washington Post* found major gaps between the views of those in rural versus urban areas. It revealed that "disagreements between rural and urban America ultimately center on fairness: who wins and loses in the new American economy, who deserves the most help in society and whether the federal government shows preferential treatment to certain types of people." Those living in rural areas face poor job prospects and feel that "government help tends to go to irresponsible people who do not deserve it."[23]

This combination of geographic, educational, and political divisions is dangerous. Those who do poorly in the transition to a digital economy are inclined to lash out at the "establishment" and the elites they feel have rigged the system. They do not trust the government to do what is right, and they believe "special interests" get all the financial benefits. These attitudes make it difficult to develop new policy solutions or redistribute resources to those who need the help. As the Brookings economist Richard Reeves has pointed out in his book *The Dream Hoarders* the class divide is strengthening in America amid fundamental changes in technology, inequality, and business models. All that makes it difficult to address important policy challenges and rethink the social contract.[24]

There are various ways to address the current mismatch in outcomes. Economic policies need to become more inclusive, both geographically and demographically. A situation in which a small number of counties generate the bulk of GDP is not tenable politically or economically.[25] If not addressed, it can be expected to lead to widespread discontent and disruption, making for a very chaotic country.[26]

As an example, an innovative program called "The Rise of the Rest," launched by the investor Stephen Case and the author J. D. Vance, seeks to encourage venture capital investment in hinterland startups as a way of stimulating greater economic activity in underserved areas. Right now, the heartland attracts only 3 percent of venture capital investment.[27] Case and Vance believe that greater attention should be paid to those regions to promote more equitable growth nationwide.

There should also be infrastructure and human capital investments in the underserved hinterlands to promote stronger development. That would leverage public resources in a way that helps underserved communities. It would balance out the economic growth that is already occurring on the coasts with better prospects elsewhere in the country. In the long run, greater investment of different kinds could boost the fortunes of areas that are lagging in economic growth and worker prosperity.

IMPROVING LEGISLATIVE REPRESENTATION

In state legislatures and the U.S. House, the mismatch problem is accentuated by inequities arising from partisan gerrymandering. In a number of states that allow their legislatures to redraw the district lines every ten years, redistricting is done in ways that advantage the majority party. Using computers and advanced scenario planning, those proposing new district lines distort representation and weaken democracy.

One example of this occurred in the state of Wisconsin. The ruling Republican Party there has engineered district lines that have given it a disproportionate share of the State Assembly seats several elections in a row. In 2012, it garnered 60 percent of the Assembly seats despite winning only 48.6 percent of the popular vote.[28] In 2014, it gained 63 percent of the seats but only 52 percent of the vote. And in 2016, it won 64 percent of the seats based on 52 percent of the popular vote.[29] Since many Democratic votes are concentrated in big cities, district lines were drawn to put those individuals in just a few districts, which makes it possible for Republicans to dominate the more numerous nonurban districts and gain an inequitable share of the seats in the legislature.

A similar problem has cropped up in the U.S. House of Representatives. An analysis of the 2016 election found that Republicans captured 49.9 percent of the popular vote nationwide but gained 55.2 percent of the seats.[30] That gave them a seat bonus and majority control even though they did not have that much support across the country.

There also are looming issues in the U.S. Senate. Since its representation is based on the principle of two seats for every state, regardless of population size, it institutionalizes a substantial disparity between political representation and economic output. With two-thirds of the senators representing areas that generate only one-third of GDP, it provides an institutional mechanism for people who are doing poorly economically to express their political discontent.

Baruch College political scientist David Birdsell anticipates this type of misrepresentation is going to increase in intensity in coming decades. He predicts that "by 2040, 70 percent of Americans will live in the 15 largest states." Based on his analysis, commentators Dionne, Ornstein, and Mann note that "70 percent of Americans [will] get all of 30 Senators and 30 percent of Americans [will] get 70 Senators."[31] Combined with the economic disparities across states, that is a recipe for populist and antigovernment movements to continue for years to come.

ABOLISHING THE ELECTORAL COLLEGE

In two elections in the past sixteen years, the winner of the presidential popular vote did not gain a majority of Electoral College votes. The reason is the same geographic problem that plagues many state legislatures and the U.S. Congress. The Electoral College allots electors in each state equal to the number of U.S. Senate and House members. That results in overrepresentation of small states and underrepresentation of large states.

The political scientists Dionne, Ornstein, and Mann make this point very clearly in their critical analysis of the Electoral College. They write, "California has 67 times more people than Wyoming does. . . . California gets one elector for every 713,637 people, Wyoming one for every 195,167. Thus, in real terms, a Wyoming voter has more than three-and-a-half times the electoral power of a California voter."[32]

These representational issues are likely to become even more problematic as the country moves into a digital economy because of the growing geographic inequities across states. Rather than being the clear exception, which was the case for most of American history, the disparities between the small number of counties that generate much of the economic activity and the larger number that are economically distressed could produce situations in which the winning presidential candidate rarely takes the popular vote. If that happened all the time, it would undermine the popular legitimacy of democracy and lead to a full-blown constitutional crisis from people's grievances over the electoral results.

To keep this from happening, we should amend the Constitution to abolish the Electoral College and choose presidents based on the popular vote. That would reduce the odds of a public backlash and put the system on firmer grounds in terms of overall legitimacy. It would eliminate a major source of popular discontent and link election outcomes more closely with public wishes.

CAMPAIGN FINANCE REFORM

Campaign finance reform is an urgent priority in the United States. Our political system has been distorted by the major parties' heavy reliance on a small number of megadonors and the corrosive impact of big money in American politics. In the 2016 elections, the top 100 donors provided $1 billion of the $1.8 billion donated to super PACs.[33] This is just one sign of a system that is overly dependent on large contributors.

The campaign finance problem is intertwined with technological innovation because many of the emerging technologies have generated tremendous wealth for a small number of individuals, and this perpetuates the financial inequality that spills over into the political system. Ultra-wealthy individuals sometimes seek to use their economic resources for political purposes and gain disproportionate influence over campaigns and governance.

To deal with this difficulty, it is important to make changes. With the reluctance of the courts to close the loopholes that have opened the door to large and secret campaign contributions, the country needs a constitutional amendment that would overturn the *Citizens United* decision, which paved the way for big donors and corporations to influence elections. Such an amendment would close the loophole that allows wealthy interests to devote hundreds of thousands of dollars to getting particular candidates elected, often in secret from the general public. That would improve the functioning of American democracy. There needs to be greater transparency in dealing with large contributions and a better way for politicians to seek office without undue dependence on wealthy individuals.

A SOLIDARITY TAX

Improving the social contract, providing worker retraining, and supporting lifetime learning are not small-scale propositions. Hundreds of billions of dollars will be needed to help workers adjust to the shifting job landscape and changing business models. Even raising income taxes on the wealthy will not come close to supporting the policy initiatives that are necessary.

For that reason, it is important to enact a 1 percent solidarity tax on net personal assets over $8 million. According to the Urban Institute, that levy would fall only on households in the top 1 percent of wealth in the United States.[34] Revenues generated by such a proposal could fund the social support necessitated by economic restructuring and technological innovation. It would cover things such as citizen accounts, lifetime learning, an expansion of the earned income tax credit, and paid family and medical leave. Not only would this possibility provide crucial funding for important programs, it would reduce the widespread inequality that has arisen in the United States.

Solidarity is an important principle for the coming era. As noted by Brian Dijkema, the program director for work and economics at Cardus, "The connections that form among us when we live and work together improve our lives and provide support."[35] In recent decades, though, these bonds have frayed and weakened people's linkages. The result has been a social contract that focuses on individual, not social responsibility, and a population that does not always want to help others. Rekindling that sense of togetherness is a pressing task for the years ahead.

CONCLUSION

The United States and the world are at a major inflection point. The increasing adoption of digital technologies and consequent changes in business models have fueled a dramatic change in the U.S. employment landscape and an increase in economic inequality.[36] Inequality in turn threatens the political process by making it difficult to address underlying social and economic issues. As noted by Rachel Nuwer in her analysis of how world civilizations fail, "Disaster comes when elites push society toward instability and eventual collapse by hoarding huge quantities of wealth and resources, and leaving little or none for commoners who vastly outnumber them."[37]

This is the problem America faces today. There is a risk of unemployment and underemployment owing to the increased use of robotics and AI, leading potentially to inequality on a massive scale. The market on its own cannot resolve this knot of issues. There is no self-corrective mechanism when economic and social inequality spring up. Historically, inequity is addressed when policies are put in place that promote social mobility and greater opportunities for all.

Failing to anticipate the serious ramifications of increasing joblessness is risky for all sides. If society and the political order are not able to help people handle the structural transformation that is coming, public anger will grow, and the populist backlash against the national and global elites will intensify.[38] More and more people will believe the system is rigged and the establishment is doing nothing to help ordinary workers.

Pushed to the extreme, middle-class voters will feel betrayed that they are not sharing in the prosperity unleashed by technological innovation, and they will experience grave anxiety about their own futures and those of their children. Those in the middle or at the bottom of the economic ladder will see prospects for themselves recede, and the political scene could turn even uglier than it is today.[39] People will look for targets and vent their frustration on perceived scapegoats. Social unrest could be widespread, and the risks to democratic rule would be substantial.[40]

However, the U.S. polity does not have to go down that path of social and political divisiveness. If we adjust our politics, social contract, and job definitions, we can deal with the coming stresses.[41] All it takes is policy and actions that encourage and make possible skills retraining, lifelong education, and a creative reimagining of the world of work. Making progress in these areas will require considerable forbearance, generosity, and far-sightedness on the part of many people. Voters will have to move in the direction of social responsibility and taking care of those who do not fare well in the digital era. They will need to look beyond individual self-interest to community benefits and shared responsibility. Even if it is a bumpy ride, there are sensible economic and political reforms that will help people navigate the treacherous terrain ahead.

NOTES

Chapter One

1 Claire Cain Miller, "The Long-Term Jobs Killer Is Not China. It's Automation," *New York Times*, December 21, 2016.

2 Eric Boehm, "Other Shoe Drops as $15 Minimum Wage Spurs Wendy's to Pursue Automated Ordering," Watchdog.org, May 13, 2016.

3 Tae Kim, "McDonald's Hits All-Time High as Wall Street Cheers Replacement of Cashiers with Kiosks," CNBC, June 22, 2017.

4 Jack Karsten and Darrell M. West, "Automation beyond the Factory," *TechTank* (blog), Brookings Institution, December 15, 2016.

5 Claire Cain Miller, "Amazon's Move Signals End of Line for Many Cashiers," *New York Times Upshot*, June 17, 2017.

6 Angel Gonzalez, "Are Amazon's Robots Job Robbers or Dance Partners?," *Seattle Times*, August 17, 2017; Sarah Kessler, "Amazon's Massive Fleet of Robots Hasn't Slowed Down Its Employment of Humans," *Quartz*, February 3, 2017.

7 Quoted in Brian Baskin, "Next Leap for Robots: Picking Out and Boxing Your Online Order," *Wall Street Journal*, July 23, 2017.

8 Will Knight, "A Robot with Its Head in the Cloud Tackles Warehouse Picking," *Technology Review*, April 5, 2017.
9 Alice Rivlin, "Seeking a Policy Response to the Robot Takeover," *Real Clear Markets*, May 2, 2017.
10 Darrell M. West, *Megachange: Economic Disruption, Political Upheaval, and Social Strife in the 21st Century* (Brookings Institution Press, 2016).
11 International Federation of Robotics, "Executive Summary World Robotics 2016 Service Robots," 2016 (https://ifr.org/downloads/press/02_2016/Executive_Summary_Service_Robots_2016.pdf).
12 James Hagerty, "Meet the New Generation of Robots for Manufacturing," *Wall Street Journal*, June 2, 2015.
13 Alison Sander and Meldon Wolfgang, "The Rise of Robotics," Boston Consulting Group, August 27, 2014 (www.bcgperspectives.com/content/articles/business_unit_strategy_innovation_rise_of_robotics/).
14 RBC Global Asset Management, "Global Megatrends: Automation in Emerging Markets," 2014 (https://us.rbcgam.com/resources/docs/pdf/whitepapers/Global_Megatrends_Automation_Whitepaper.pdf).
15 Jennifer Smith, "A Robot Can Be a Warehouse Worker's Best Friend," *Wall Street Journal*, August 3, 2017.
16 Anonynous CEO, quoted in Moises Naim, "As Robots Take Our Jobs, Guaranteed Income Might Ease the Pain," *Huffington Post*, July 18, 2016.
17 Kim Tingley, "Learning to Love Our Robot Co-Workers," *New York Times Magazine*, February 23, 2017.
18 Quoted in Baskin, "Next Leap for Robots."
19 Grace Lordan and David Neumark, "People versus Machines: The Impact of Minimum Wages on Automatable Jobs," NBER Working Paper 26337 (Cambridge, Mass.: National Bureau of Economic Research, August 2017). Also see Chico Harlan, "Rise of the Machines," *Washington Post*, August 5, 2017.
20 Quoted in Alana Semuels, "Robots Will Transform Fast Food," *The Atlantic*, December 7, 2017.
21 John Markoff, "Korean Team Wins Pentagon's Crisis Robotics Contest," *New York Times*, June 8, 2015.
22 Gary Shteyngart, "Thinking outside the Bots," *Smithsonian*, June 2017, pp. 66–80.
23 Prudence Ho and Jason Gale, "A Case of Chicken vs. Machine," *Bloomberg Businessweek*, January 16–22, 2017, p. 18.
24 Dexter Roberts and Rachel Chang, "China's Robot Revolution," *Bloomberg Businessweek*, May 1–7, 2017, pp. 32–34.
25 Keith Bradsher, "A Robot Revolution in China as Car Manufacturers Look to Cut Costs," *New York Times*, May 13, 2017.

26 Conner Forrest, "Chinese Factory Replaces 90% of Humans with Robots, Production Soars," *Tech Republic*, July 30, 2015.

27 Nick Statt, "iPhone Manufacturer Foxconn Plans to Replace Almost Every Human Worker with Robots," *The Verge*, December 30, 2016.

28 Beatrice Gitau, "Smart Hotel: Japan Opens a Hotel Run by Robots," *Christian Science Monitor*, July 18, 2015.

29 Donna St. George, "Peyton's Awesome Virtual Self, a Robot That Allows Girl with Cancer to Attend School," *Washington Post*, November 28, 2016.

30 Eitan Wilf, "Sociable Robots, Jazz Music, and Divination: Contingency as a Cultural Resource for Negotiating Problems of Intentionality," *American Ethnologist: Journal of the American Ethnological Society*, November 6, 2013, p. 605 (http://onlinelibrary.wiley.com/doi/10.1111/amet.12041/abstract).

31 Mike Murphy, "Amazon Tests out Robots That Might One Day Replace Warehouse Workers," *Quartz*, June 1, 2015; Gregory Wallace, "Amazon Deploys Army of Robots in Gen Warehouses," *CNN Money*, December 1, 2014.

32 Michael Belfiore, "Delivery Robot," *Bloomberg Businessweek*, May 23–29, 2016, p. 33.

33 Quoted in Dan Zak, "I Am Your New Robot Security Guard," *Washington Post*, September 26, 2017.

34 Cynthia Breazeal, "The Personal Side of Robots," speech, SXSW (South by Southwest), Austin, Tex., March 13, 2015.

35 Nick Leiber, "Europe Bets on Robots to Help Care for Seniors," *Bloomberg Businessweek*, March 21–27, 2016, p. 38.

36 John Markoff, "As Aging Population Grows, So Do Robotic Health Aides," *New York Times*, December 4, 2015.

37 Felix Gillette, "Baby's First Virtual Assistant," *Bloomberg Businessweek*, January 3, 2017.

38 Rachel Botsman, "Co-Parenting with Alexa," *New York Times*, October 8, 2017, p. 5.

39 Thi-Hai-Ha Dang and Adriana Tapus, "Stress Game: The Role of Motivational Robotic Assistance in Reducing User's Task Stress," *International Journal of Social Robotics*, April 2015.

40 Jenny Kleeman, "The Race to Build the World's First Sex Robot," *The Guardian*, April 27, 2017.

41 Julie Beck, "Who's Sweating the Sexbots?," *The Atlantic*, September 30, 2015; Caitlin Gibson, "The Future of Sex Includes Robots and Holograms," *Washington Post*, January 14, 2016.

42 George Gurley, "Is This the Dawn of the Sexbots?," *Vanity Fair*, May, 2015.

43 Kleeman, "The Race to Build the World's First Sex Robot."

44 Ibid.

45 Ibid.

46 Ibid.

47 Alyson Krueger, "Future Sex Is Here," *New York Times*, October 29, 2017.

48 Quoted in Daniel Weisfield, "Peter Thiel at Yale," MBA blog, Yale School of Management, April 27, 2013 (https://dev-som.yale.edu/blog/peter-thiel-at-yale-we-wanted-flying-cars-instead-we-got-140-characters?blog=3490).

49 Ibid.

50 European Parliament, "European Civil Law Rules in Robotics," 2016 (www.europarl.europa.eu/RegData/etudes/STUD/2016/571379/IPOL_STU(2016)571379_EN.pdf).

Chapter Two

1 Quoted in Adi Robertson, "Treasury Secretary 'Not at All' Worried about Robots Taking Jobs," *The Verge*, March 24, 2017.

2 Quoted in Ina Fried, "Elon Musk: 'There Will Not Be a Steering Wheel' in 20 Years," *Axios*, July 15, 2017.

3 Shukla S. Shubhendu and Jaiswal Vijay, "Applicability of Artificial Intelligence in Different Fields of Life," *International Journal of Scientific Engineering and Research* 1, no. 1 (2013), pp. 2347–78.

4 Jenna Wortham, "Silicon Valley Has Fallen in Love with Chatbots," *New York Times Magazine*, April 24, 2016.

5 Executive Office of the President, "Artificial Intelligence, Automation, and the Economy," White House, December 2016 (https://obamawhitehouse.archives.gov/sites/whitehouse.gov/files/documents/Artificial-Intelligence-Automation-Economy.PDF), and "Preparing for the Future of Artificial Intelligence," White House, October 2016 (https://obamawhitehouse.archives.gov/sites/default/files/whitehouse_files/microsites/ostp/NSTC/preparing_for_the_future_of_ai.pdf).

6 Thomas Davenport, Jeff Loucks, and David Schatsky, "Bullish on the Business Value of Cognitive" (Deloitte, 2017), p. 3 (www2.deloitte.com/us/en/pages/deloitte-analytics/articles/cognitive-technology-adoption-survey.html).

7 Shubhendu and Vijay, "Applicability of Artificial Intelligence in Different Fields of Life."

8 Luke Dormehl, *Thinking Machines: The Quest for Artificial Intelligence—and Where It's Taking Us Next* (New York: Penguin-TarcherPerigee, 2017).

9 Shubhendu and Vijay, "Applicability of Artificial Intelligence in Different Fields of Life."

10 Michael Lewis, *Flash Boys: A Wall Street Revolt* (New York: Norton, 2015).

11 Cade Metz, "In Quantum Computing Race, Yale Professors Battle Tech Giants," *New York Times*, November 14, 2017, p. B3.
12 Andrei A. Kirilenko and Andrew W. Lo, "Moore's Law versus Murphy's Law: Algorithmic Trading and Its Discontents," *Journal of Economic Perspectives* 27, no. 2 (2013), pp. 51–72.
13 Christian Davenport, "Future Wars May Depend as Much on Algorithms as on Ammunition, Report Says," *Washington Post*, December 3, 2017.
14 Ibid.
15 Kevin Desouza, Rashmi Krishnamurthy, and Gregory Dawson, "Learning from Public Sector Experimentation with Artificial Intelligence," *TechTank* (blog), Brookings Institution, June 23, 2017.
16 Cecille De Jesus, "AI Lawyer 'Ross' Has Been Hired by Its First Official Law Firm," *Futurism*, May 11, 2016.
17 Paul Mozur, "China Sets Goal to Lead in Artificial Intelligence," *New York Times*, July 21, 2017, p. B1.
18 Paul Mozur and John Markoff, "Is China Outsmarting American Artificial Intelligence?," *New York Times*, May 28, 2017.
19 "China May Match or Beat America in AI," *Economist*, July 15, 2017.
20 Paul Mozur and Keith Bradsher, "China's A.I. Advances Help Its Tech Industry, and State Security," *New York Times*, December 3, 2017.
21 Simon Denyer, "China's Watchful Eye," *Washington Post*, January 7, 2018.
22 Dominic Barton, Jonathan Woetzel, Jeongmin Seong, and Qinzheng Tian, "Artificial Intelligence: Implications for China" (New York: McKinsey Global Institute, April 2017), p. 1.
23 Ibid., p. 7.
24 Andrew McAfee and Erik Brynjolfsson, *Machine Platform Crowd: Harnessing Our Digital Future* (New York: Norton, 2017).
25 Ibid.
26 Armand Joulin and Tomas Mikolov, "Inferring Algorithmic Patterns with Stack-Augmented Recurrent Nets," *ArKiv*, June 1, 2015.
27 Nathaniel Popper, "Stocks and Bots," *New York Times Magazine*, February 28, 2016.
28 Ibid.
29 Ibid.
30 Ibid.
31 Pat Regnier, "Coming for Your Trading Desk," *Bloomberg Businessweek*, June 26, 2017, pp. 22–23.
32 Rasmus Rothe, "Applying Deep Learning to Real-World Problems," *Medium*, May 23, 2017.
33 Ibid.

34 Ray Kurzweil, "Integrated Circuits," *New York Times Book Review*, March 19, 2017, p. 13.

35 Cameron Kerry and Jack Karsten, "Gauging Investment in Self-Driving Cars," Brookings Institution, October 16, 2017.

36 Portions of this section are drawn from Darrell M. West, "Driverless Cars in China, Europe, Japan, Korea, and the United States," Brookings Institution, September, 2016.

37 Interview with experts of Baidu, July 14, 2016.

38 Waymo, "On the Road to Fully Self-Driving," Waymo Safety Report, 2017 (https://storage.googleapis.com/sdc-prod/v1/safety-report/waymo-safety-report-2017-10.pdf).

39 Test results come from the open database Labeled Faces in the Wild (LFW), which can be found at http://vis-www.cs.umass.edu/lfw/results.html#attsim.

40 Interview with experts of Baidu, July 12, 2016.

41 Ibid.

42 Yuming Ge, Xiaoman Liu, Libo Tang, and Darrell M. West, "Smart Transportation in China and the United States," Center for Technology Innovation, Brookings Institution, December, 2017.

43 Peter Holley, "Uber Signs Deal to Buy 24,000 Autonomous Vehicles from Volvo," *Washington Post*, November 20, 2017.

44 Farhad Manjoo, "Think Amazon's Drone Delivery Idea Is a Gimmick? Think Again," *New York Times*, August 10, 2016.

45 Ibid.

46 Ibid.

47 Kaya Yurieff, "Amazon Patent Reveals Drone Delivery 'Beehives,'" *CNN Tech*, June 23, 2017.

48 Anousha Sakoui, "Can VR Find a Seat in the Parlor?," *Bloomberg Businessweek*, May 29–June 4, 2017, p. 22.

49 Rachel Metz, "Augmented Reality Is Finally Getting Real, *Technology Review*, August 2, 2012.

50 Cade Metz, "A New Way for Therapists to Get Inside Heads: Virtual Reality," *New York Times*, July 30, 2017.

51 Laurits Christensen, Wes Marcik, Greg Rafert, and Carletta Wong, "The Global Economic Impacts Associated with Virtual and Augmented Reality," Analysis Group, 2016 (www.analysisgroup.com/uploaded-files/content/news_and_events/news/analysis_group_vr_economic_impact_executive_summary.pdf).

52 Joshua Kopstein, "The Dark Side of VR," *The Intercept*, December 23, 2016.

53 For more information on this, see Darrell M. West, "The Ethical Dilem-

mas of Virtual Reality," *TechTank* (blog), Brookings Institution, April 18, 2016.

54 Joshua Brustein and Spencer Soper, "Who's Alexa?," *Bloomberg Businessweek*, May 2–8, 2016, pp. 31–33.

55 Penelope Green, "'Alexa, Where Have You Been All My Life?,'" *New York Times*, July 11, 2017.

56 Kevin Desouza and Rashmi Krishnamurthy, "Chatbots Move Public Sector toward Artificial Intelligence," *TechTank* (blog), Brookings Institution, June 2, 2017.

57 Dieter Bohn, "The Machine Is Learning," *The Verge*, May 17, 2017.

58 Ibid.

59 Mike Isaac, "Facebook Bets on Bots for Its Messenger App," *New York Times*, April 12, 2016.

60 Jane Levere, "A.I. May Book Your Next Trip (with a Human Assist)," *New York Times*, May 30, 2016.

61 Yongdong Wang, "Your Next New Best Friend Might Be a Robot," *Nautilus*, September 14, 2017.

62 Osonde Osoba and William Welser IV, "The Risks of Artificial Intelligence to Security and the Future of Work" (Santa Monica, Calif.: RAND Corp., December 2017) (www.rand.org/pubs/perspectives/PE237.html).

63 Elaine Glusac, "As Airbnb Grows, So Do Claims of Discrimination," *New York Times*, June 21, 2016.

64 "Joy Buolamwini," *Bloomberg Businessweek*, July 3, 2017, p. 80.

65 Ian Tucker, "'A White Mask Worked Better': Why Algorithms Are Not Colour Blind," *The Guardian*, May 28, 2017.

66 Jessica Guynn, "Palantir Charged with Discriminating against Asians," *USA Today*, September 26, 2016; Jessica Guynn, "Palantir Settles Asian Hiring Discrimination Lawsuit," *USA Today*, April 25, 2017.

67 Jon Valant, "Integrating Charter Schools and Choice-Based Education Systems," *Brown Center Chalkboard* (blog), Brookings Institution, June 23, 2017.

68 Levi Tillemann and Colin McCormick, "Roadmapping a U.S.-German Agenda for Artificial Intelligence Policy," New America Foundation, March 2017, p. 4.

69 Katie Benner, "Airbnb Vows to Fight Racism, But Its Users Can't Sue to Prompt Fairness," *New York Times*, June 19, 2016.

70 John Quain, "Cars Suck up Data about You. Where Does It All Go?," *New York Times*, July 27, 2017.

71 Jeff Asher and Rob Arthur, "Inside the Algorithm That Tries to Predict Gun Violence in Chicago," *New York Times Upshot*, June 13, 2017.

72 Caleb Watney, "It's Time for Our Justice System to Embrace Artificial Intelligence," *TechTank* (blog), Brookings Institution, July 20, 2017.
73 Asher and Arthur, "Inside the Algorithm That Tries to Predict Gun Violence in Chicago."
74 Tucker, "'A White Mask Worked Better.'"
75 Cliff Kuang, "Can A.I. Be Taught to Explain Itself?," *New York Times Magazine*, November 21, 2017.
76 Oren Etzioni, "How to Regulate Artificial Intelligence," *New York Times*, September 1, 2017.
77 "Ethical Considerations in Artificial Intelligence and Autonomous Systems," unpublished paper, IEEE Global Initiative, 2017.
78 Ritesh Noothigattu, Snehalkumar Gaikwad, Edmond Awad, Sohan Dsouza, Iyad Rahwan, Pradeep Ravikumar, and Ariel Procaccia, "A Voting-Based System for Ethical Decision Making," *Computers and Society*, MIT Media Lab, September 20, 2017 (www.media.mit.edu/publications/a-voting-based-system-for-ethical-decision-making/).
79 Joseph Aoun, *Robot-Proof: Higher Education in the Age of Artificial Intelligence* (MIT Press, 2017).
80 Danielle Paquette, "Her Dilemma: Do I Let My Employer Microchip Me?," *Washington Post*, July 25, 2017.
81 Executive Office of the President, "Artificial Intelligence, Automation, and the Economy" and "Preparing for the Future of Artificial Intelligence."
82 Eric Siegel, "Predictive Analytics Interview Series: Andrew Burt," *Predictive Analytics Times*, June 14, 2017.

Chapter Three

1 *Quote Investigator*, "How Will You Get Robots to Pay Union Dues?," June 9, 2017.
2 Ian King, "5G Networks Will Do Much More than Stream Better Cat Videos," *Bloomberg News*, May 2, 2016.
3 Portions of this chapter are drawn from Darrell M. West, "How 5G Enables the Health Internet of Things," Brookings Institution, July 14, 2016.
4 Interview with Asha Keddy, June 7, 2016.
5 Tom Peters, "FCC Workshop Reveals Secrets of 5G," blog post, Hogan Lovells, March 15, 2016.
6 Mark Scott, "What 5G Will Mean for You," *New York Times*, February 21, 2016.
7 Numbers cited in Carrie MacGillivray, "The Internet of Things Is Poised to Change Everything, Says IDC," *Business Wire*, October 3, 2013; Charles McLellan, "The Internet of Things and Big Data," *ZDNet*, March 2, 2015.

8 The device number comes from King, "5G Networks Will Do Much More than Stream Better Cat Videos," while the sensor number was provided by Bridget Karlin in a June 10, 2016, interview.
9 Hadley Weinzierl, "Digital Universe Invaded by Sensors," EMC.com, April 9, 2014.
10 Scott, "What 5G Will Mean for You."
11 Marc Andreessen, "Why Software Is Eating the World," *Wall Street Journal*, August 20, 2011.
12 Quoted in Sean Buckley, "AT&T Will Launch SDN Service in 63 Countries Simultaneously This Year," press release, AT&T, May 23, 2016.
13 Rajat Sahni, "New Report Study SDN/NFV Technologies: Innovative Use Cases and Operator Strategies," *Industry Today*, April 11, 2016.
14 David Goldman, "What Is 5G?," *CNN Money*, December 4, 2015.
15 Tadilo Endeshaw Bogale and Long Bao Le, "Massive MIMO and Millimeter Wave for 5G Wireless HetNet: Potentials and Challenges," *IEEE Vehicular Technology Magazine*, October 21, 2015.
16 Ibid.
17 Quoted in Jonathan Rockoff, "Remote Patient Monitoring Lets Doctors Spot Trouble Early," *Wall Street Journal*, February 16, 2015.
18 "Cellular Technologies Enabling the Internet of Things," 4G Americas, November, 2015 (www.5gamericas.org/files/6014/4683/4670/4G_Americas_Cellular_Technologies_Enabling_the_IoT_White_Paper_-_November_2015.pdf).
19 Ian Scales, "How Much Is Being Spent on IoT, and in What Sectors?," *Telecom TV*, June 24, 2016.
20 Robert Hume and Jeff Looney, "Telemedicine and Facility Design," *HFM Magazine*, February, 2016.
21 Chii-Wann Lin and others, "Taipei Citizen Telecare Service System for Hypertension Management in Elders," in *Proceedings of the 2014 Annual SRII Global Conference* (Washington, D.C.: IEEE Computer Society, 2014), pp. 157–80.
22 Dana Wollman, "The Internet of Toddlers," *Engadget*, January 7, 2014.
23 White House, "President Obama's Precision Medicine Initiative," press release, January 30, 2015 (https://obamawhitehouse.archives.gov/the-press-office/2015/01/30/fact-sheet-president-obama-s-precision-medicine-initiative).
24 Hye-Jung Chun and others, "Second-Generation Sequencing for Cancer Genome Analysis," *Cancer Genomics*, 2014.
25 Eric Dishman, "Getting to the Next Step with Personalized Medicine," *Intel Blog*, February 25, 2016.

26 National Institutes of Health, Precision Medicine Initiative Cohort Program (www.nimhd.nih.gov/programs/collab/pmi/).
27 Jocelyn Kaiser, "NIH's 1-Million-Volunteer Precision Medicine Study Announces First Pilot Projects," *Science,* February 25, 2016.
28 Jessica Davis, "Penn Medicine's Modern Big Data Initiative's Applications Alert Doctors of At-Risk Patients," *Information Week*, October 6, 2015.
29 Jonathan Vanian, "Intel's Cancer Cloud Gets New Recruits," *Fortune*, March 31, 2016.
30 Interview of Bob Rogers, June 6, 2016.
31 "The Internet of Things and Healthcare Policy Principles," white paper, Intel, undated (www.intel.com/content/dam/www/public/us/en/documents/white-papers/iot-healthcare-policy-principles-paper.pdf).
32 Sandeep Vashist, Peter Luppa, Leslie Yeo, Aydogan Ozcan, and John Luong, "Emerging Technologies for Next-Generation Point-of-Care Testing," *Trends in Biotechnology* 33, no. 11 (2015), pp. 692–705.
33 University of Virginia Health System, "Home-Based Coordinated Care Management," unpublished paper, undated.
34 Alan Snell and Julia Smalley, "Beacon Community Research Study: Reducing Hospital Readmissions via Remote Patient Management," Indiana Health Information Exchange, 2013.
35 Care Innovations, "How Mississippi Is Leading the Way in Innovation," 2015.
36 Paul Budde Communication, "Global Digital Economy: E-Health and M-Health. Insights, Stats and Analysis," August 12, 2015, p. 2.
37 Ibid., p. 17.
38 Andras Petho, David Fallis, and Dan Keating, "ShotSpotter Detection System Documents 39,000 Shooting Incidents in the District," *Washington Post*, November 2, 2013.
39 Portions of this section come from Darrell M. West and Dan Bernstein, "Benefits and Best Practices of Public Safe City Innovation," Brookings Institution, 2017.
40 Tjerk Timan, "The Body-Worn Camera as a Transitional Technology," *Surveillance & Society* 14, no. 1 (2016), pp. 145–49.
41 Barak Ariel, William Farrar, and Alex Sutherland, "The Effect of Police Body-Worn Cameras on Use of Force and Citizens' Complaints against the Police," *Journal of Quantitative Criminology* 31, no. 3 (September 2015), 509–35.
42 Angela Godwin, "Advanced Metering Infrastructure: Drivers and Benefits in the Water Industry," *Water World*, undated (www.waterworld.com/

articles/print/volume-27/issue-8/editorial-features/special-section-advanced-metering-infrastructure/advanced-metering-infrastructure-drivers-and-benefits-in-the-water-industry.html).

43 Portions of this section come from Darrell M. West, "Driverless Cars in China, Europe, Japan, Korea, and the United States," Center for Technology Innovation, Brookings Institution, September 2016.

44 Pacific Institute, "Metering in California," September 2014.

45 "Smarter Water Management: Parks, Recreation and Open Spaces," fact sheet, Miami-Dade County and IBM Technology Projects, undated.

46 Danielle Bochove, "A More Automated Gold Mine," *Bloomberg Businessweek*, October 30, 2017, pp. 26–27.

47 Portions of this section come from West, "Driverless Cars in China, Europe, Japan, Korea, and the United States."

48 Li Shufu, "Paving the Way for Autonomous Cars in China," *Wall Street Journal*, April 21, 2016.

49 Chris Buckley, "Beijing's Electric Bikes, the Wheels of E-Commerce, Face Traffic Backlash," *New York Times*, May 30, 2016.

50 Interview with experts of Baidu, July 12, 2016.

51 James Anderson, Nidhi Kalra, Karlyn Stanley, Paul Sorensen, Constantine Samaras, and Oluwatobi Oluwatola, "Autonomous Vehicle Technology: A Guide for Policymakers" (Santa Monica, Calif.: RAND Corp., 2016), p. xvi.

52 Tatiana Schlossberg, "Stuck in Traffic, Polluting the Inside of Our Cars," *New York Times*, August 29, 2016.

53 Daniel Fagnant and Kara Kockelman, "The Travel and Environmental Implications of Shared Autonomous Vehicles Using Agent-Based Model Scenarios," *Transportation Research Part C* 40 (2014), pp. 1–13.

54 Daniel Shoup, "Cruising for Parking," *Access* 30 (2007), pp. 16–22.

55 Bruce Weindelt, "Digital Transformation of Industries: Automotive Industry," World Economic Forum in collaboration with Accenture, January 2016, p. 4.

56 Harald Bauer, Ondrej Burkacky, and Christian Knochenhauer, "Security in the Internet of Things" (New York: McKinsey, May 2017).

Chapter Four

1 Edward Bellamy, *Looking Backward: 2000–1887* (Boston: Houghton-Mifflin, 1888).

2 Dawn Nakagawa, "The Second Machine Age Is Approaching," *Huffington Post*, February 24, 2015.

3 The number of total employees comes from Jerry Davis, "Capital Markets and Job Creation in the 21st Century," Center for Effective Public Management, Brookings Institution, December 30, 2015, p. 7. The market capitalization numbers for General Motors are based on a stock price of $53.50 per share as noted in the *Chicago Tribune*, "G.M. Stock Weak in Irregularly Higher Market," March 3, 1962, and 283,488,664 outstanding shares in common stock, as noted in General Motors' 1963 10-K statement filed with the Securities and Exchange Commission, p. 25. The market capitalization value for AT&T is based on a stock price of $106 per share and 243,062,000 outstanding shares in common stock as noted the *New York Times*, "AT&T Earnings Set Marks in '62," January 3, 1963. I used the U.S. Bureau of Labor Statistics CPI inflator to convert 1962 to 2017 dollars.
4 Mary Meeker, "Internet Trends," Kleiner Perkins, 2017.
5 Robert Gebeloff and Karl Russell, "How the Growth of E-Commerce Is Shifting Retail Jobs," *New York Times*, July 7, 2017, p. B1.
6 U.S. Bureau of Labor Statistics, "Current Population Survey" for prime-age men, 1948–2017. See also Executive Office of the President, "The Long-Term Decline in Prime-Age Male Labor Force Participation," White House, June 2016 (https://obamawhitehouse.archives.gov/sites/default/files/page/files/20160620_cea_primeage_male_lfp.pdf).
7 U.S. Bureau of Labor Statistics, "Labor Force Participation," September 2016.
8 Executive Office of the President, "The Long-Term Decline in Prime-Age Male Labor Force Participation," p. 3.
9 Eleanor Krause and Isabel Sawhill, "What We Know and Don't Know about Declining Labor Force Participation," Center on Children and Families, Brookings Institution, May 17, 2017.
10 Cited in Michael Schuman, "Why Wages Aren't Growing," *Bloomberg Businessweek*, September 25, 2017.
11 David Rotman, "Who Will Own the Robots," *MIT Technology Review*, September 2015.
12 Martin Ford, *The Lights in the Tunnel: Automation, Accelerating Technology, and the Economy of the Future* (CreateSpace, 2009), p. 237. See also Martin Ford's more recent book, *Rise of the Robots: Technology and the Threat of a Jobless Future* (New York: Basic Books, 2015).
13 Katja Grace, John Salvatier, Allan Dafoe, Baobao Zhang, and Owain Evans, "When Will AI Exceed Human Performance? Evidence from AI Experts," arXiv.org, May 30, 2017.
14 "Manufacturing under the Trump Administration," Sixth Annual John

Hazen White Forum on Public Policy, Brookings Institution, July 13, 2017.

15 U.S. Bureau of Labor Statistics, "Employment Projections: 2014–2024 Summary," December 8, 2015.

16 Ibid.

17 Quoted in Harold Meyerson, "Technology and Trade Policy Is Pointing America toward a Job Apocalypse," *Washington Post*, March 26, 2014. The original paper is Carl Benedict Frey and Michael Osborne, "The Future of Employment: How Susceptible Are Jobs to Computerisation?," faculty paper, Oxford University, September 17, 2013.

18 Frey and Osborne, "The Future of Employment," pp. 57–72.

19 Jeremy Bowles, "Chart of the Week: 54% of EU Jobs at Risk of Computerisation," blog post, Bruegel.org, July 24, 2014. See also Georgios Petropoulos, "Do We Understand the Impact of Artificial Intelligence on Employment?," blog post, Bruegel.org, April 27, 2017.

20 Ben Schiller, "How Soon before Your Job Is Done by a Robot?," *Fast Coexist*, January 6, 2016.

21 James Manyika, Michael Chui, Mehdi Miremadi, Jacques Bughin, Katy George, Paul Willmott, and Martin Dewhurst, "A Future That Works: Automation, Employment, and Productivity" (New York: McKinsey Global Institute, January 2017).

22 James Manyika, Susan Lund, Michael Chui, Jacques Bughin, Jonathan Woetzel, Parul Batra, Ryan Ko, and Saurabh Sanghui, "Jobs Lost, Jobs Gained: Workforce Transitions in a Time of Automation" (New York: McKinsey Global Institute, December 2017).

23 Aaron Smith and Janna Anderson, "AI, Robotics, and the Future of Jobs," Pew Research Center, August 6, 2014.

24 Melanie Arntz, Terry Gregory, and Ulrich Zierahn, "The Risk of Automation for Jobs in OECD Countries: A Comparative Analysis," Organization for Economic Cooperation and Development, Working Paper 189 (OECD, 2016), p. 4.

25 Erik Brynjolfsson and Andrew McAfee, *The Second Machine Age: Work, Progress, and Prosperity in a Time of Brilliant Technologies* (New York: Norton, 2014), p. 11.

26 Daron Acemoglu and Pascual Restrepo, "Robots and Jobs: Evidence from US Labor Markets," NBER Working Paper 23285 (Cambridge, Mass.: National Bureau of Economic Research, March 2017), abstract. Also see Claire Cain Miller, "Evidence That Robots Are Winning the Race for American Jobs," *New York Times*, March 29, 2017, p. B3.

27 Lawrence Summers, "The Economic Challenge of the Future: Jobs," *Wall Street Journal*, July 7, 2014.

28 Quoted in Christopher Matthews, "Summers: Automation Is the Middle Class'[s] Worst Enemy," *Axios,* June 4, 2017.
29 Quoted in David Rotman, "How Technology Is Destroying Jobs," *MIT Technology Review*, June 12, 2013 (www.technologyreview.com/featured-story/515926/how-technology-is-destroying-jobs/).
30 Quoted in Rotman, "How Technology Is Destroying Jobs."
31 Mark Muro, Sifan Liu, Jacob Whiton, and Siddharth Kulkarni, "Digitalization and the American Workforce," Metropolitan Policy Program, Brookings Institution, November 2017.
32 Quoted in Melissa Kearney, Brad Hershbein, and David Boddy, "The Future of Work in the Age of the Machine," The Hamilton Project, Brookings Institution, February 2015.
33 Ruchir Sharma, "Robots Won't Kill the Workforce: They'll Save the Economy," *Washington Post*, December 4, 2016.
34 Quoted in Kia Kokalitcheva, "Self-Driving Cars Will Boost the Job Market," *Axios AM*, May 31, 2017.
35 Aaron Smith, "Public Predictions for the Future of Workforce Automation," Pew Research Center, March 10, 2016.
36 Aaron Smith, "U.S. Views of Technology and the Future," Pew Research Center, April 2014 (www.pewinternet.org/files/2014/04/US-Views-of-Technology-and-the-Future.pdf).
37 Aaron Smith, "Automation in Everyday Life," Pew Research Center, October 4, 2017 (www.pewinternet.org/2017/10/04/automation-in-everyday-life/).
38 "Making It in America: The View from America," Burson-Marsteller and PSB survey, June 2017, unpublished report, pp. 25, 29.
39 Aaron Smith and Monica Anderson, "Automation in Everyday Life," Pew Research Center, October 4, 2017.
40 Ibid., p. 26.
41 Ibid.
42 Nathan Bomey, "Automation Puts Jobs in Peril," *USA Today*, February 6, 2017.
43 A. T. Kearney, "Adapting to Disruption," 2017, p. 5.
44 United Kingdom Commission for Employment and Skills, "The Future of Work: Jobs and Skills in 2030," February 2014 (www.gov.uk/government/publications/jobs-and-skills-in-2030).
45 Bureau of Labor Statistics, Current Population Survey, Household Data, June 2017 (www.bls.gov/opub/mlr/2016/article/labor-force-participation-what-has-happened-since-the-peak.htm).
46 Costanza Biavaschi, Werner Eichhorst, Corrado Giulietti, Michael Kendzia, Alexander Muravyev, Janneke Pieters, Nurai Rodriguez-Planas,

Ricarda Schmidl, and Klaus Zimmermann, "Youth Unemployment and Vocational Training," World Development Report, World Bank, 2013.
47 U.S. Department of Education, "Science, Technology, Engineering and Math," 2014 (www.ed.gov/sites/default/files/stem-overview.pdf).
48 Jeffrey Sachs, "Smart Machines and the Future of Jobs," *Boston Globe*, October 10, 2016.
49 Ibid.
50 World Bank, "Unemployment, Youth Total," 2017 (https://data.worldbank.org/indicator/SL.UEM.1524.ZS).
51 Richard Adler and Rajiv Mehta. "Catalyzing Technology to Support Family Caregiving," National Alliance for Caregiving, 2014 (www.caregiving.org/wp-content/uploads/2010/01/Catalyzing-Technology-to-Support-Family-Caregiving_FINAL.pdf).
52 Matthew Clark, Jongil Lim, Girma Tewolde, and Jaerock Kwon, "Affordable Remote Health Monitoring System for the Elderly Using Smart Mobile Devices," *Sensors & Transducers* 184, no.1 (January 31, 2015), pp. 77–83.
53 Laura Robinson, Sheila R. Cotten, Hiroshi Ono, Anabel Quan-Haase, Gustavo Mesch, Wenhong Chen, Jeremy Schultz, Timothy M. Hale, and Michael J. Stern, "Digital Inequalities and Why They Matter," *Information, Communication & Society* 18, no. 5 (2015), pp. 569–82.
54 David Weil, *The Fissured Workplace* (Harvard University Press, 2014).
55 Karen Shook, "Review of 'The Fissured Workplace,'" *Times Higher Education*, March 6, 2014.
56 Niam Yaraghi and Shamika Ravi, "The Current and Future State of the Sharing Economy," Brookings Institution, December 29, 2016.
57 Matt Sinclair, "5 Questions for . . . Zoe Baird," *Philanthropy News Digest*, September 12, 2016.
58 Danny Vinik, "The Real Future of Work," *Politico*, January/February, 2018, pp. 80–87.
59 Ian Hathaway and Mark Muro, "Ridesharing Hits Hyper-Growth," *The Avenue* (blog), Brookings Institution, June 1, 2017.
60 Adam Minter, "China Is the Future of the Sharing Economy," *Bloomberg News*, May 18, 2017.
61 Amanda Erickson, "A Chinese Umbrella-Sharing Start-up Just Lost Nearly All of Its 300,000 Umbrellas," *Washington Post*, July 12, 2017.
62 Amy Qin, "In China, Umbrellas and Basketballs Join the Sharing Economy," *New York Times*, May 28, 2017.
63 Liz Alderman, "Feeling 'Pressure All the Time' on Europe's Treadmill of Temporary Work," *New York Times*, February 9, 2017.
64 Yaraghi and Ravi, "The Current and Future State of the Sharing Economy."

65 Eli Lehrer, "The Future of Work," *National Affairs*, Summer 2016, p. 36.
66 Ellen Huet, "The Humans Hiding behind the Chatbots," *Bloomberg Businessweek*, May 9–15, 2016, pp. 34–35.
67 Jamie Horsley, "Backgrounder on China's Sharing Economy," June 2017, unpublished paper.
68 Noam Scheiber, "Uber Has a Union of Sorts, but Faces Doubts on Its Autonomy," *New York Times*, May 12, 2017.
69 Shift: The Commission on Work, Workers, and Technology, "Report of Findings," Shift Commission.Work, 2017 (www.newamerica.org/new-america/policy-papers/shift-commission-report-findings/). (Shift is a joint iniative of New America and Bloomberg.)
70 Griffith Insurance Education Foundation, "Millennial Generation Attitudes about Work and the Insurance Industry," February 6, 2012 (www.theinstitutes.org/doc/Millennial-Generation-Survey-Report.pdf).
71 Lindsey Pollack, "Attitudes and Attributes of Millennials in the Workplace," Deloitte, September 12, 2014.
72 Job Centre Plus, "Volunteering while Getting Benefits" (London: UK Department for Work and Pensions, October 2010) (www.gov.uk/government/uploads/system/uploads/attachment_data/file/264508/dwp1023.pdf).
73 Quoted in Derek Thompson, "A World without Work," *The Atlantic*, July/August, 2015.
74 Melinda Sandler Morill and Sabrina Wulff Pabilonia, "What Effects Do Macroeconomic Conditions Have on Families' Time Together?," Leibniz Information Centre for Economics, 2012 (http://hdl.handle.net/10419/58561).
75 National Endowment for the Arts, "Arts Data Profile," August 2016.
76 National Endowment for the Arts, "Surprising Findings in Three New NEA Reports on the Arts," January 12, 2015.
77 Ibid.
78 Christopher Ingraham, "Poetry Is Going Extinct, Government Data Show," *Washington Post*, April 24, 2015.
79 Nielsen, "Year-End Music Report," January 9, 2017.
80 Nancy Vogt, "Audio: Fact Sheet," Pew Research Center, June 15, 2016.
81 Craft Yarn Council, "Knitting and Crocheting Are Hot," January 2015.
82 Centers for Disease Control and Prevention, National Center for Health Statistics, "Prevalence of Obesity among Adults and Youth," November 2015.
83 Statista, "Total Number of Memberships at Fitness Centers/Health Clubs in the U.S. from 2000 to 2015," 2016.

84 U.S. Department of Health and Human Services, "Leisure-Time Physical Activity," May 2016.

85 PHIT America, "America's 15 Fastest Growing Sports and Activities," May 5, 2015.

86 Marlynn Wei, "New Survey Reveals the Rapid Rise of Yoga," *Harvard Health Blog*, March 7, 2016.

Chapter Five

1 Quoted in Kia Kokalitcheva, "Self-Driving Cars Will Boost the Job Market," *Axios AM*, May 31, 2017.

2 Nicolas Colin and Bruno Palier, "Social Policy for a Digital Age," *Foreign Affairs* 94, no. 4 (July/August 2015), pp. 29–33.

3 Kaiser Family Foundation, "Health Insurance Coverage of the Total Population," database, July 10, 2017 (www.kff.org/other/state-indicator/total-population/?currentTimeframe=0&sortModel=%7B%22colId%22:%22Location%22,%22sort%22:%22asc%22%7D).

4 Henry Alford, "The Tricky Etiquette of Co-Working Spaces," *New York Times*, November 5, 2016.

5 Colin Bradford and Roger Burkhardt, "Empowering People to Control Their Futures," Policy Report, Brookings Institution, March 9, 2017, unpublished paper.

6 Eli Lehrer, "The Future of Work," *National Affairs*, Summer 2016, p. 48.

7 Jon Greenberg, "'Medicaid Expansion Drove Health Insurance Coverage under Health Law,' Rand Paul Says," *Politifact,* January 15, 2017.

8 Seth Harris and Alan Krueger, "A Proposal for Modernizing Labor Laws for Twenty-First-Century Work: The 'Independent Worker,'" The Hamilton Project, Brookings Institution, December 2015, p. 3.

9 Ibid., p. 6.

10 Daniel Araya and Sunil Johal, "Work and Social Policy in the Age of Artificial Intelligence," *TechTank* (blog), Brookings Institution, February 28, 2017.

11 Laura Addati, Naomi Cassirer, and Katherine Gilchrist, *Maternity and Paternity at Work: Law and Practice across the World* (Geneva: International Labour Organization, 2014).

12 AEI-Brookings Working Group on Paid Family Leave ,"Paid Family and Medical Leave: An Issue Whose Time Has Come," American Enterprise Institute-Brookings Institution, May 2017, p. 2 (www.brookings.edu/wp-content/uploads/2017/06/es_20170606_paidfamilyleave.pdf).

13 Ibid.
14 Erik Brynjolfsson and Andrew McAfee, *The Second Machine Age: Work, Progress, and Prosperity in a Time of Brilliant Technologies* (New York: W. W. Norton, 2014), pp. 238–39.
15 "The Tax Policy Briefing Book," Tax Policy Center, Urban Institute and Brookings Institution, 2016 (www.taxpolicycenter.org/briefing-book/key-elements/family/eitc.cfm).
16 Ibid.
17 Cass Sunstein, "A Poverty-Buster That's No Liberal Fantasy," *Bloomberg View*, August 13, 2015.
18 Elizabeth Kneebone and Natalie Holmes, "Strategies to Strengthen the Earned Income Tax Credit," Brookings Institution, December 9, 2015.
19 Natalie Holmes and Alan Berube, "The Earned Income Tax Credit and Community Economic Stability," Brookings Institution, November 20, 2015.
20 Alan Berube, "Want to Help the Working Class? Pay the EITC Differently," *The Avenue* (blog), Brookings Institution, June 28, 2017.
21 Steve Holt, "Periodic Payment of the Earned Income Tax Credit Revisited," Policy Report, Brookings Institution, December 2015, p. 17.
22 "What Is Trade Adjustment Assistance?" (U.S. Department of Labor, July 14, 2017).
23 Tom DiChristopher, "Sizing Up the Trade Adjustment Assistance Program," CNBC, June 26, 2015.
24 Kara Reynolds and John Palatucci, "Does Trade Adjustment Assistance Make a Difference?," American University, August 2008, unpublished paper.
25 DiChristopher, "Sizing Up the Trade Adjustment Assistance Program."
26 Mark Muro, "Failure to Adjust: The Case of Auto-IRA," *The Avenue* (blog), Brookings Institution, May 8, 2017.
27 Timothy Martin, "The Champions of the 401(k) Lament the Revolution They Started," *Wall Street Journal*, January 2, 2017.
28 Thomas Heath, "The 401(k) Match Is Back, and It's Getting Bigger," *Washington Post*, July 18, 2017.
29 Quoted in Colin and Palier, "Social Policy for a Digital Age." See also Scott Santens, "Everything You Think You Know about the History and Future of Jobs Is Wrong," Institute for Ethics and Emerging Technologies, August 19, 2015.
30 Ben Schiller, "A Universal Basic Income Is the Bipartisan Solution to Poverty We've Been Waiting For," *Fast Coexist*, March 16, 2015.

31 Robert Skidelsky, "Minimum Wage or Living Income," *Project Syndicate*, July 16, 2015.
32 Max Ehrenfreund, "The Issue That Could Unite Conservatives and Socialists," *Washington Post*, June 7, 2016.
33 Peter Goodman, "Free Money for the Jobless," *New York Times*, December 18, 2016.
34 Anthony Painter and Chris Thoung, "Creative Citizen, Creative State: The Principled and Pragmatic Case for a Universal Basic Income," *Medium*, 2015, p. 11.
35 Rob Atkinson, "13 Things to Know about How Automation Impacts Jobs," *Huffington Post*, May 10, 2017.
36 Ibid.
37 Charles Kenny, "Give Poor People Cash," *The Atlantic*, September 25, 2015.
38 Derek Thompson, "A World without Work," *The Atlantic*, July/August, 2015.
39 Greg Beach, "Finland Prepares Universal Basic Income Experiment," *Inhabitat*, November 4, 2015.
40 Tafline Laylin, "Dutch City to Hand Out Free Basic Income in New Social Experiment," *Inhabitat*, June 30, 2015.
41 Libby Brooks, "Universal Basic Income Trials Being Considered in Scotland," *The Guardian*, January 1, 2017.
42 Quoted in Andrew Flowers, "What Would Happen If We Just Gave People Money?," *Five Thirty Eight*, April 25, 2016.
43 Helena Bachmann, "Swiss Say 'No' to a Guaranteed Income from the Government," *USA Today*, June 6, 2016.
44 Lukas Golder and others, "Real Public Debate on Unconditional Basic Income," GFS Bern, June 5, 2016.
45 Belinda Tasker, "Could a Basic Income Protect You from the Rise of Robots?," *West Australian*, September 23, 2017.
46 Cited in Eli Lehrer, "The Future of Work," *National Affairs*, Summer 2016, p. 44.
47 Claire Cain Miller, "How to Beat the Robots," *New York Times*, March 7, 2017.
48 Ellen Huet, "The Humans Hiding behind the Chatbots," *Bloomberg Businessweek*, May 9–15, 2016, pp. 34–35.
49 Kathryn Dill, "Job-Stealing Robots Should Pay Income Taxes," CNBC, February 17, 2017.
50 Lawrence Summers, "Picking on Robots Won't Deal with Job Destruction," *Washington Post*, March 5, 2017.
51 Noah Smith, "What's Wrong with Bill Gates' Robot Tax," *Bloomberg News*, February 28, 2017.

52 James Stewart, "Tax Reform for the Wealthy: Lower Rates but Lose Breaks," *New York Times*, September 22, 2017, p. B1.

53 Michelle Fox, "Why We Need a Global Wealth Tax," CNBC, March 10, 2015.

54 Daphne Chen, Fatih Guvenen, Gueorgui Kambourov, and Burhanettin Kuruscu, "Efficiency Gains from Wealth Taxation," February 15, 2013.

55 Adam Nossiter, "Emmanuel Macron's Unwanted New Title: 'President of the Rich,'" *New York Times*, November 1, 2017.

56 "Wealth Tax," *Wikipedia* (accessed May 23, 2017).

57 "Trump Proposes Massive One-Time Tax on the Rich," CNN, November 9, 1999.

58 Urban Institute, "Nine Charts about Wealth Inequality in America," Urban.org, 2015 (last update October 5) (http://apps.urban.org/features/wealth-inequality-charts/).

59 Robert Frank, "The Top 1% of Americans Now Control 38% of the Wealth," CNBC, September 28, 2017. See also "Trends in Family Wealth, 1989 to 2013" (Congressional Budget Office, August 2016), p. 2; and Emmanuel Saez and Gabriel Zucman, "Wealth Inequality in the United States since 1913: Evidence from Capitalized Income Tax Data," *Quarterly Journal of Economics*, May 2016, pp. 519–78.

60 The 2017 number comes from U.S. Federal Reserve Bank, "Recent Developments in Household Net Worth and Domestic Nonfinancial Debt," 2017 (www.federalreserve.gov/releases/z1/current/z1.pdf). For historical data, see the CBO's August 2016 "Trends in Family Wealth, 1989 to 2013," p. 1.

61 Quoted in Anna Bernasek, "Looking beyond Income, to a Tax on Wealth," *New York Times*, February 9, 2013.

62 Max Ehrenfreund, "Trump's Proposals Could Hike Taxes for Nearly a Quarter of the Middle Class," *Washington Post*, July 12, 2017. Also see William Gale, Surachai Khitatrakun, and Aaron Krupkin, "Winners and Losers after Paying for the Tax Cuts and Jobs Act," Tax Policy Center, December 8, 2017.

63 Chye-Ching Huang, "Corporate Tax Cuts Could Hurt—Not Help—Workers," *Off the Charts* (blog) (Washington, D.C.: Center on Budget and Policy Priorities, July 20, 2017).

Chapter Six

1 Daniel Araya and Heather McGowan, "Education and Accelerated Change," *Brown Center Chalkboard* (blog), Brookings Institution, September 14, 2016.

2 Monte Whaley, "Colorado Students Find Niche in Tech and Hands-On School Programs," *Denver Post*, December 25, 2016.

3 Elizabeth Mann, "Connecting Community Colleges with Employers: A Toolkit for Building Successful Partnerships," Brown Center on Education Policy, Brookings Institution, July 2017.

4 John Donovan and Cathy Benko, "AT&T's Talent Overhaul," *Harvard Business Review*, October 2016.

5 Ibid.

6 Harry Holzer, "Will Robots Make Job Training (and Workers) Obsolete? Workforce Development in an Automating Labor Market," Policy Report, Brookings Institution, June 19, 2017, p. 5.

7 "73% of Adults Consider Themselves Lifelong Learners," Pew Research Center, March 22, 2016.

8 Ibid.

9 Eric Hanushek and Ludger Woessmann, "Apprenticeship Programs in a Changing Economic World," *Brown Center Chalkboard* (blog), Brookings Institution, June 28, 2017.

10 Portions of this section come from Darrell M. West, *Digital Schools: How Technology Can Transform Schools* (Brookings Institution Press, 2012).

11 Elaine Allen and Jeff Seaman, "Class Differences: Online Education in the United States, 2010" (Boston: Babson Survey Research Group, 2010), p. 5.

12 Basmat Parsad and Laurie Lewis, "Distance Education at Degree-Granting Postsecondary Institutions" (National Center for Education Statistics, U.S. Department of Education, 2008).

13 Elaine Allen and Jeff Seaman, "Digital Learning Compass: Distance Education Enrollment Report" (Boston: Babson Survey Research Group, 2017).

14 Barbara Means, Yukie Toyama, Robert Murphy, Marianne Bakia, and Karla Jones, "Evaluation of Evidence-Based Practices in Online Learning: A Meta-Analysis and Review of Online Learning Studies" (U.S. Department of Education, Office of Planning, Evaluation, and Policy Development, September 2010).

15 Kurt Eisele-Dyrli, "Mobile Goes Mainstream," *District Administration*, February 2011 (www.districtadministration.com/article/mobile-goes-mainstream).

16 Garry Falloon, "Making the Connection: Moore's Theory of Transactional Distance and Its Relevance to the Use of a Virtual Classroom in Postgraduate Online Teacher Education," *Journal of Research on Technology in Education* 43, no. 3 (2011), pp. 187–209.

17 Daphne Koller, "Online Education for the 21st Century," faculty paper, Stanford University, undated.
18 Quoted in Bryant Urstadt, "The Math of Khan," *Bloomberg Business Week*, May 23–29, 2011, p. 76.
19 Quoted in Dawn Nakagawa, "The Second Machine Age Is Approaching," *Huffington Post*, February 24, 2015.
20 Ibid.
21 United Kingdom Commission for Employment and Skills, "The Future of Work: Jobs and Skills in 2030," February 2014, p. 106 (www.gov.uk/government/publications/jobs-and-skills-in-2030).
22 Thomas Arnett, "Teaching Is Ripe for Machine Assistance," Teach for America, June 5, 2017 (www.teachforamerica.org/one-day-magazine/teaching-ripe-machine-assistance).
23 "Innovate to Educate: System [Re]Design for Personalized Learning. A Report from the 2010 Symposium" (Washington, D.C.: Software and Information Industry Association, 2010), p. 8.
24 Portions of this section come from West, *Digital Schools: How Technology Can Transform Schools.*
25 Howard Gardner, *Frames of Mind: The Theory of Multiple Intelligences* (New York: Basic Books, 1983).
26 Ibid.
27 John Dewey, *Schools of Tomorrow* (New York: Dutton, 1915), p. 18.
28 "Innovate to Educate: System [Re]Design for Personalized Learning," p. 18.
29 Ibid., p. 19.
30 See school description at www.hightechhigh.org.
31 Ibid.
32 John Hechinger, "A Virtual Education," *Bloomberg Business Week*, June 6–12, 2011, p. 77.
33 Quoted in Trip Gabriel, "Speaking Up in Class, Silently, Using the Tools of Social Media," *New York Times*, May 13, 2011, p. A1.
34 Ted Kolderie and Tim McDonald, "How Information Technology Can Enable 21st Century Schools" (Washington, D.C.: Information Technology and Innovation Foundation, July 2009), p. 2.
35 Kemal Derviş, "A New Birth for Social Democracy," op-ed, Project Syndicate, Brookings Institution, June 10, 2015.
36 Aspen Institute Future of Work, "Lifelong Learning and Training Accounts," 2018.
37 White House, "President's Plan to Provide Americans with Job Training and Employment Services," March 12, 2012.
38 Peter McClure, "Grubstake," *Change*, June 1976, p. 41.

39 Kelli Grant, "Michigan Joins Ranks of Schools with Free Tuition," CNBC, June 16, 2017.
40 Kelli Grant, "If You Can't Get New York's Free Tuition, Here Are 10 More States with Cheap College Costs," CNBC, May 17, 2017.
41 Derviş, "A New Birth for Social Democracy."

Chapter Seven

1 John Green, "Scary New World," *New York Times*, November 7, 2008.
2 John Micklethwait, Megan Murphy, and Ellen Pollock, "Don't Gamble, Invest," *Bloomberg Businessweek*, June 13, 2016, p. 47.
3 Portions of this section come from Darrell M. West, *Megachange: Economic Disruption, Political Upheaval, and Social Strife in the 21st Century* (Brookings Institution Press, 2016).
4 Cited in Tyler Cowen, "Industrial Revolution Comparisons Aren't Comforting," *Bloomberg News*, February 16, 2017.
5 David Bornstein, "The Art of Getting Opponents to 'We,'" *New York Times*, November 3, 2015.
6 Tom Wheeler, "Did Technology Kill the Truth?," Brookings Institution, November 14, 2017.
7 Portions of this section come from Darrell M. West, *Billionaires: Reflections on the Upper Crust* (Brookings Institution Press, 2014).
8 Thomas Piketty and Emmanuel Saez, "Income Inequality in the United States, 1913–1998," *Quarterly Journal of Economics* 118 (2003), pp. 1–39. For 1999 to 2008 numbers, see the web page of Emmanuel Saez (http://emlab.berkeley.edu/users/saez). See also Richard Burkhauser and others, "Recent Trends in Top Income Shares in the USA: Reconciling Estimates from March CPS and IRS Tax Return Data," NBER Working Paper 15320 (Cambridge, Mass.: National Bureau of Economic Research, September 2009); and Thomas Piketty, *Capital in the Twenty-First Century* (Harvard University Press, 2014).
9 The 2012 income numbers are from Emmanuel Saez, "Striking It Richer: The Evolution of Top Incomes in the United States," faculty paper, Department of Economics, University of California, Berkeley, September 3, 2013 (http://elsa.berkeley.edu/~saez/saez-UStopincomes2012.pdf).
10 Ed Harris and Frank Sammartino, "Trends in the Distribution of Household Income, 1979–2009" (Congressional Budget Office, August 6, 2012).
11 Piketty, *Capital in the Twenty-First Century.*
12 Colin Bradford, *Reframing Globalization toward Better Social Outcomes* (Berlin: Friedrich-Ebert-Stiftung, May 2017), p. 3.

13 Benjamin Page, Larry Bartels, and Jason Seawright, "Democracy and the Policy Preferences of Wealthy Americans," *Perspectives on Politics* 11 (March 2013), pp. 51–73.

14 Ibid., p. 55.

15 Kay Lehman Schlozman, Sidney Verba, and Henry Brady, *The Unheavenly Chorus: Unequal Political Voice and the Broken Promise of American Democracy* (Princeton University Press, 2012).

16 Lee Drutman, "The 1,000 Donors Most Likely to Benefit from McCutcheon—and What They Are Most Likely to Do" (Sunlight Foundation, October 2, 2013).

17 Page, Bartels, and Seawright, "Democracy and the Policy Preferences of Wealthy Americans," pp. 53–54.

18 "American National Election Study 2012 Preliminary Release Codebook," University of Michigan, June 13, 2013, p. 1039.

19 Page, Bartels, and Seawright, "Democracy and the Policy Preferences of Wealthy Americans."

20 Ibid.

21 *Global Wealth Databook: 2012* (Credit Suisse Research Institute, October 2012), p. 127 (http://piketty.pse.ens.fr/files/Davies%20et%20al%20 2012_global_wealth_databook.pdf).

22 Page, Bartels, and Seawright, "Democracy and the Policy Preferences of Wealthy Americans."

23 Martin Gilens, *Affluence and Influence: Economic Inequality and Political Power in America* (Princeton University Press, 2012).

24 Martin Gilens and Benjamin Page, "Testing Theories of American Politics: Elites, Interest Groups, and Average Citizens," *Perspectives on Politics* 12, no. 3. (Fall 2014), p. 2.

25 Nicolas Colin and Bruno Palier, "Social Policy for a Digital Age," *Foreign Affairs*, July/August 2015.

26 Quoted in Nathan Gardels, "French Reforms Aim for a New Social Contract in the Age of Disruption," *Washington Post*, December 1, 2017.

27 Liz Alderman, "In Sweden, Happiness in a Shorter Workday Can't Overcome the Cost," *New York Times*, January 6, 2017.

28 Kevin Desouza, "Autonomous Vehicles Will Cost Local Governments Big Bucks," *Slate*, June 16, 2015.

29 Jack Ewing, "Robocalype Now? Central Bankers Argue Whether Automation Will Kill Jobs," *New York Times*, June 28, 2017.

30 David Leonhardt, "The American Dream, Quantified at Last," *New York Times*, December 11, 2016, p. 2.

31 William Galston, *Anti-Pluralism: The Populist Threat to Liberal Democracy* (Yale University Press, forthcoming).
32 Philip Bump, "Places That Saw More Job Loss to Robots Were Less Likely to Support Hillary Clinton," *Washington Post*, March 29, 2017.
33 "Full Text of Mark Zuckerberg's Harvard Graduation Speech," *USA Today*, May 25, 2017.
34 Mark Muro and Sifan Liu, "Another Clinton-Trump Divide: High-Output America vs. Low-Output America," *The Avenue* (blog), Brookings Institution, November 29, 2016.
35 Kim Hart, "The Large Parts of America Left behind by Today's Economy," *Axios*, September 25, 2017 (www.axios.com/americas-fractured-economic-well-being-2488460340.html).
36 Ben Casselman, "A Start-Up Slump Is a Drag on the Economy," *New York Times*, September 20, 2017.
37 Franklin Foer, *World without Mind: The Existential Threat of Big Tech* (New York: Penguin Press, 2017).
38 Steve LeVine, "Artificial Intelligence Pioneer Calls for the Breakup of Big Tech," *Axios*, Sepember 20, 2017.
39 Seth London and Bradley Tusk, "How to Save the Rust Belt," *Politico*, September 6, 2017; Andrew Ross Sorkin, "From Bezos to Walton, Big Investors Back Fund for 'Flyover' Start-Ups," *New York Times*, December 4, 2017.
40 Quoted in Sam Wetherell, "Richard Florida Is Sorry," *Jacobin*, August 19, 2017.
41 Jim Tankersley, "Donald Trump Lost Most of the American Economy in This Election," *Washington Post*, November 22, 2016.
42 Ibid.
43 Portions of this section are drawn from Darrell M. West, "How to Combat Fake News and Disinformation," Brookings Insitutution, December 2017.
44 Jen Weedon, William Nuland, and Alex Stamos, "Information Operations," *Facebook*, April 27, 2017.
45 Craig Silverman, "This Analysis Shows How Viral Fake Election News Stories Outperformed Real News on Facebook," *BuzzFeedNews*, November 16, 2016.
46 Craig Timberg and Elizabeth Dwoskin, "Russian Content on Facebook, Google and Twitter Reached Far More Users than Companies First Disclosed, Congressional Testimony Says," *Washington Post*, October 30, 2017.
47 Tim Wu, "Did Twitter Kill the First Amendment?," *New York Times*, October 28, 2017, p. A9.

48 Marc Fisher, John Cox, and Peter Hermann, "Pizzagate: From Rumor, to Hashtag, to Gunfire in D.C.," *Washington Post*, December 6, 2016.
49 Ibid.
50 Quoted in Craig Silverman and Jeremy Singer-Vine, "Most Americans Who See Fake News Believe It, New Survey Says," *BuzzFeed News*, December 6, 2016.
51 Hunt Allcott and Matthew Gentzkow, "Social Media and Fake News in the 2016 Election," NBER Working Paper 23089 (Cambridge, Mass.: National Bureau of Economic Research, April 2017), p. 4.
52 Emilio Ferrara, Onur Varol, Clayton Davis, Filippo Menczer, and Alessandro Flammini, "The Rise of Social Bots," *Communications of the ACM* 59, no. 7 (July 2016), pp. 96–104.
53 Ibid.
54 Michela Del Vicario, Alessandro Bessi, Fabiana Zollo, Fabio Petroni, Antonio Scala, Guido Caldarelli, Eugene Stanley, and Walter Quattrociocchi, "The Spreading of Misinformation Online," *PNAS* 113, no. 3 (2016), pp. 554–59.
55 David Lazer, Matthew Baum, Nir Grinberg, Lisa Friedland, Kenneth Joseph, Will Hobbs, and Carolina Mattsson, "Combating Fake News: An Agenda for Research and Action" (Harvard Shorenstein Center on Media, Politics and Public Policy and Harvard Ash Center for Democratic Governance and Innovation, May 2017), p. 5.
56 Quoted in Belinda Goldsmith, "Trust the News? Most People Don't, Social Media Even More Suspect," Reuters, June 21, 2017.
57 Martin Gilens and Benjamin Page, "Testing Theories of American Politics: Elites, Interest Groups, and Average Citizens," *Perspectives on Politics* 12, no. 3 (Fall 2014), pp. 564–81.
58 Eduardo Porter, "What's at Stake in a Health Bill That Slashes the Safety Net," *New York Times*, March 21, 2017.

Chapter Eight

1 Steelah Kolhatkar, "Welcoming Our New Robot Overlords," *New Yorker*, October 23, 2017.
2 AEI-Brookings Working Group on Paid Family Leave, "Paid Family and Medical Leave: An Issue Whose Time Has Come," American Enterprise Institute-Brookings Institution, May 2017 (https://www.brookings.edu/wp-content/uploads/2017/06/es_20170606_paidfamilyleave.pdf).
3 Colin Bradford and Roger Burkhardt, "Empowering People to Control Their Futures," Policy Report, Brookings Institution, March 9, 2017; Eli

Lehrer, "The Future of Work," *National Affairs*, Summer 2016, p. 48 (www.nationalaffairs.com/publications/detail/the-future-of-work).

4 Anne Case and Angus Deaton, "Mortality and Morbidity in the 21st Century," Brookings Panel on Economic Activity, May 1, 2017.

5 Carol Graham, Sergio Pinto, and John Juneau, "The Geography of Desperation in America," Brookings Institution, July 24, 2017.

6 Carol Graham, "The Unhappiness of the US Working Class," op-ed, Brookings Institution, July 10, 2017.

7 Brian Fuller, "Building Better Preschools—but for Which Kids?," *Brown Center Chalkboard* (blog), Brookings Institution, July 20, 2017.

8 Satya Nadella, *Hit Refresh* (New York: HarperCollins, 2017).

9 Tim Craig and Nicole Lewis, "As Opioid Overdoses Exact a Higher Price, Communities Ponder Who Should Be Saved," *Washington Post*, July 15, 2017.

10 Robert Gordon, "The Political Pendulum Will Swing Back," *Axios*, July 16, 2017.

11 Daron Acemoglu and Simon Johnson, "It's Time to Found a New Republic," *Foreign Policy*, August 15, 2017.

12 Benjamin Page and Martin Gilens, *Democracy in America? What Has Gone Wrong and What We Can Do About It* (University of Chicago Press, 2018).

13 William Galston and Clara Hendrickson, "A Policy at Peace with Itself: Antitrust Remedies for Our Concentrated, Uncompetitive Economy," Policy Report, Brookings Institution, January 5, 2018.

14 Tom Wheeler, "Did Technology Kill the Truth?," Brookings Institution, November 14, 2017.

15 Clara Hendrickson and William Galston, "Automation Presents a Political Challenge, but Also an Opportunity," *TechTalk* (blog), Brookings Institution, May 18, 2017.

16 Lawrence Summers, "The Economic Challenge of the Future: Jobs," *Wall Street Journal*, July 7, 2014; Christopher Matthews, "Summers: Automation Is the Middle Class' Worst Enemy," *Axios*, June 4, 2017.

17 E. J. Dionne, Norman Ornstein, and Thomas Mann, *One Nation after Trump* (New York: St. Martin's Press, 2017), p. 5.

18 Portions of this section are drawn from Darrell M. West, *Megachange: Economic Disruption, Political Upheaval, and Social Strife in the 21st Century* (Brookings Institution Press, 2016).

19 William Galston, "Telling Americans to Vote, or Else," *New York Times*, November 5, 2011.

20 International Institute for Democratic Electoral Assistance, "Compulsory Voting," undated (www.idea.int/vt/compulsory_voting.cfm#practicing).

21 William Galston and E. J. Dionne, "The Case for Universal Voting: Why

Making Voting a Duty Would Enhance Our Elections and Improve Our Government," Center for Effective Public Management, Brookings Institution, September 2015, p. 4.

22 J. D. Vance, *Hillbilly Elegy: A Memoir of a Family and Culture in Crisis* (New York: HarperCollins, 2016).

23 Jose DelReal and Scott Clement, "Rural Divide," *Washington Post*, June 17, 2017.

24 Richard Reeves, *The Dream Hoarders* (Brookings Institution Press, 2017).

25 Mark Muro and Sifan Liu, "Another Clinton-Trump Divide: High-Output America vs. Low-Output America," *The Avenue* (blog), Brookings Institution, November 29, 2016.

26 Amy Liu, "To Create Economic Opportunities, Cities Must Confront Their Past—and Look to the Future," *The Avenue* (blog), Brookings Institution, July 17, 2017.

27 Jon Swartz and Jessica Guynn, "JD Vance, Steve Case Want the Heartland's Start-Up Pitches," *USA Today*, June 26, 2017.

28 Adam Liptak and Michael Shear, "Supreme Court Hears 'Good Evidence' Voting Maps Entrenched a Party in Power, Justice Says," *New York Times*, October 3, 2017.

29 Mark Stern, "Partisan Gerrymandering Got the Sotomayor Treatment," *Slate*, October 4, 2017.

30 Molly Reynolds, "Republicans in Congtress Got a 'Seats Bonus' This Election (Again)," *FixGov* (blog), Brookings Institution, November 22, 2016.

31 Quoted in Dionne, Ornstein, and Mann, *One Nation after Trump*, p. 30.

32 Dionne, Ornstein, and Mann, *One Nation after Trump*, p. 29.

33 Paul Blumenthal, "Super PAC Mega-Donors Expand Election Influence with Record $1 Billion In Contributions," *Huffington Post*, December 19, 2016.

34 Urban Institute, "Nine Charts about Wealth Inequality in America," Urban.org, 2015 (last update October 5) (http://apps.urban.org/features/wealth-inequality-charts/).

35 Brian Dijkema, "Reviving Solidarity," *National Affairs* 34 (Winter 2018), p. 135.

36 West, *Megachange*.

37 Rachel Nuwer, "How Western Civilization Could Collapse," *BBC News*, April 18, 2017.

38 Vance, *Hillbilly Elegy*.

39 Reeves, *The Dream Hoarders*.

40 William Galston, *Anti-Pluralism: The Populist Threat to Liberal Democracy* (Yale University Press, forthcoming).

41 West, *Megachange*.

INDEX